The LogicPrep Guide to Advanced Math

LogicPrep

About this book

This book is the culmination of many hours of diligent work by the following people: Jesse Kolber, Roger Reiersen, Jamie Kenyon, Helen Moss, Brian Siberine, Molly Pickel, Matthew Kwong, Chad Schwam, and Alex Wurm. Our dynamic team collectively has over 30 years in the test-prep industry and 20,000 hours of preparing students for both the ACT and SAT.

Published by LogicPrep Tutoring, 2016
ISBN: 978-0-9851060-5-8

Copyright Notice
© LogicPrep Tutoring. All Rights Reserved.
Copying and electronic transmission are strictly forbidden, including, but not limited to, e-mail, facsimile, CD-ROM, DVD-ROM, tape, or any medium for the purpose of distribution. Printing and distribution of this publication is strictly prohibited without the authors' expressed written permission.

Contents

A Note from LogicPrep 1
SohCahToa Drill 3
Non-Right Triangle Trig Drill 5
All-Angle Trigonometry 8
Converting Angles Drill 23
Unit Circle Drill 27
Unit Circle 29
Graphing Trig Functions Drill 41
Trig Identities Drill 47
Trigonometry Graphing and IDs 50
Polynomial Algebra 57
Completing the Square 72
Even and Odd Functions 84
Conics 87
Ellipses 97
Logs Drill 111
Logs 113
Sequences 123
Combinations and Permutations 129
Imaginary Numbers 137
Matrices 141
Advanced Mixed Question Set 1 149
Advanced Mixed Question Set 2 157
Advanced Mixed Question Set 3 165
Advanced Mixed Question Set 4 183

A Note from LogicPrep

Hi there. Welcome to the high-stakes, much-dreaded, your-whole-life-seems-to-hang-in-the-balance world of the ACT.

Here's the good news: If you're reading this book, you're already smart – not because you chose our book (well, not just because) but because you've realized that the best way to get your ACT score where you want it is to prepare. Prepare logically, even.

To be sure, your ACT score will be only a part of your entire college application. Your grades, extracurriculars, outside activities and volunteer work, and of course your main essay and supplements will all factor in, but let's be honest: your ACT scores will be a big part of your application.

The Math section is scored out of 36 – which is the highest score – and is averaged together with the English, Reading, and Science sections to form a Composite Score, again out of 36. You have to learn how the ACT applies and tests your knowledge.

We're not in the room with you, but we will offer our wisdom honed from years of working with all different kinds of students, so we can alert you to common (and sometimes not-so-common) mistakes… and some variations on test-taking strategy. Now after you've done the hard work of learning and mastering the skills, we also need to make sure you have mastered three more skills: be confident, be calm, and be careful.

1. Be confident – you have learned the knowledge you need and seen how the ACT tests it so be strong: ACTs don't vary much from test to test.
2. Be calm – becoming emotional on this test will work against you. Whether you feel you are performing well or not, you must stay calm instead of becoming excited or nervous.
3. Be careful – even experienced test-takers can be sloppy, or can misread a question, or forget to plug an answer choice into the whole sentence… and so let themselves get caught off guard by a well-written but incorrect answer.

The road to success may be a little bit different for everyone – and the exact final destinations will vary too. But here's the promise we make:

If you work hard to learn the skills we teach, and you practice and revise your test taking skills, then your score will improve, often by a lot. How much improvement you attain depends on you: on your hard work and on your commitment. Lastly, we do teach these skill sets so you can do great on the ACT – but we wouldn't be in this business if we did not also believe we were teaching you things that will be vital for the rest of your lives.

So jump in. The water's fine. Have some fun and know you are making an investment that will pay off – on the ACT and beyond!

-The LogicPrep Team

SohCahToa
Quick Drill

1. $\sin B =$

2. $\cos B =$

3. $\tan B =$

4. $\sec B =$

5. $\csc B =$

6. $\cot B =$

7. $\arcsin\left(\dfrac{a}{c}\right) =$

8. $\cos^{-1}\left(\dfrac{a}{c}\right) =$

9. $\tan^{-1}\left(\dfrac{a}{b}\right) =$

10. $\operatorname{arcsec}\left(\dfrac{c}{b}\right) =$

11. $\csc^{-1}\left(\dfrac{c}{b}\right) =$

12. $\operatorname{arccot}\left(\dfrac{b}{a}\right) =$

SohCahToa
Quick Drill

Answer Key

#	Answer
1	$\dfrac{b}{c}$
2	$\dfrac{a}{c}$
3	$\dfrac{c}{b}$
4	$\dfrac{a}{c}$
5	$\dfrac{c}{b}$
6	$\dfrac{a}{b}$
7	A
8	B
9	A
10	A
11	B
12	A

Non-Right Triangle Trig
Quick Drill

1. State the law of cosines:

2. State the law of sines:

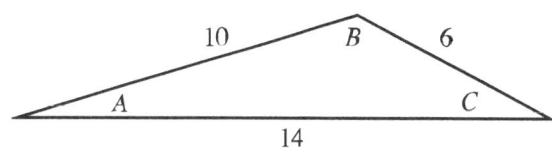

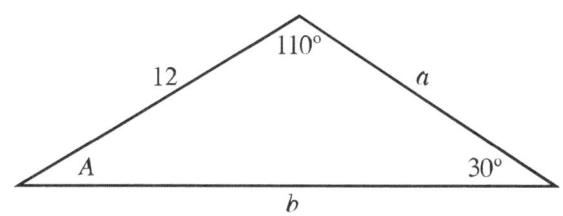

3. Find b:

5. Find B:

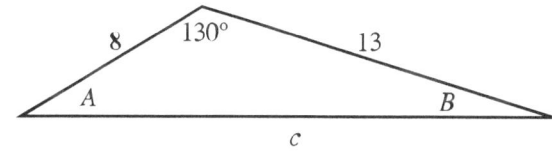

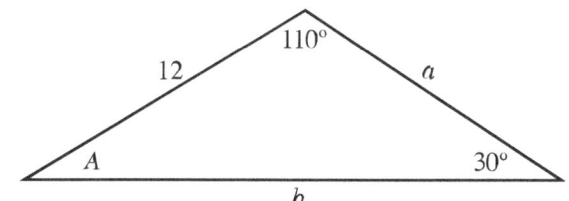

4. Find a:

6. Find c:

Non-Right Triangle Trig
Quick Drill

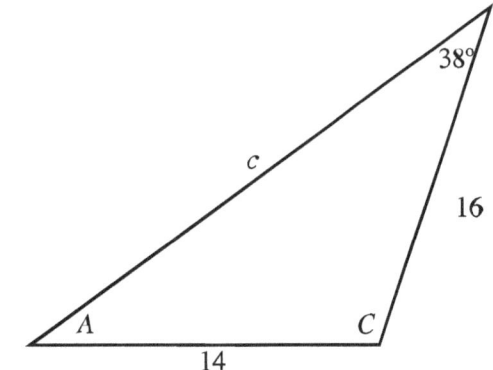

7. Find A:

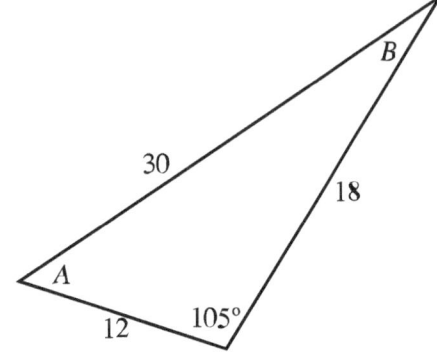

9. Find B:

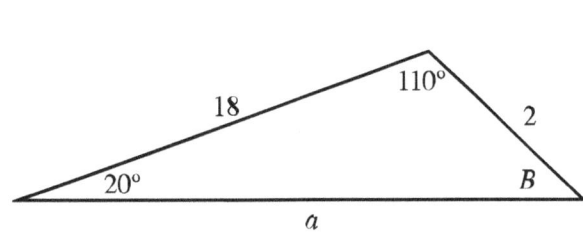

8. Find a:

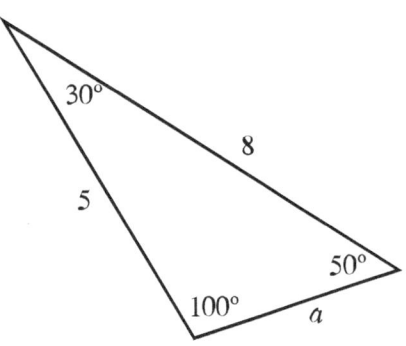

10. Find a:

Non-Right Triangle Trig
Quick Drill

Answer Key

#	Answer
1	$c^2 = a^2 + b^2 - 2ab\cos C$
2	$\dfrac{a}{\sin A} = \dfrac{b}{\sin B} = \dfrac{c}{\sin C}$
3	22.6
4	15.4
5	120°
6	19.1
7	44.7°
8	18.8
9	22.7°
10	4.4

All-Angles Trigonometry
Advanced Problem Set 1

1. In the triangle below, $\sin a =$

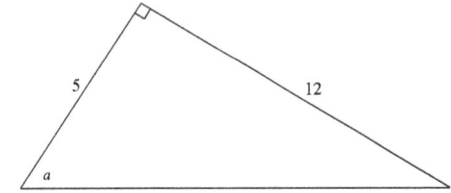

   A. $\dfrac{5}{24}$

   B. $\dfrac{5}{13}$

   C. $\dfrac{5}{12}$

   D. $\dfrac{5}{6}$

   E. $\dfrac{12}{13}$

2. In in the figure below, $\overline{BC} = 5.3$ in, $x = 57°$, and $y = 71°$. What is the length of $\overline{AB}$?

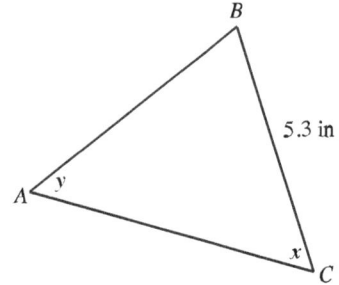

   F. 0.17

   G. 0.21

   H. 3.17

   J. 4.70

   K. 5.98

All-Angles Trigonometry
Advanced Problem Set 1

3. If $\csc x = \dfrac{1}{\sin x}$, then which of the following trigonometric equations is equivalent to $f(x) = \cos x \csc x$?

   A. $g(x) = \sin x$

   B. $g(x) = \cot x$

   C. $g(x) = \sec x$

   D. $g(x) = \cos x$

   E. $g(x) = \tan x$

4. In trigonometry, an angle of $\dfrac{-5\pi}{4}$ radians has the same sine and cosine and an angle that has which of the following degrees of measure?

   F. $45°$

   G. $90°$

   H. $135°$

   J. $180°$

   K. $225°$

All-Angles Trigonometry
Advanced Problem Set 1

5. The lengths of the sides of the triangle shown below are given in feet. Which of the following equations gives the degree measure θ?
(Note: For any triangle, $c^2 = a^2 + b^2 - 2ab\cos C$, where a, b, and c are the lengths of the sides opposite angles with measures A, B, and C, respectively.)

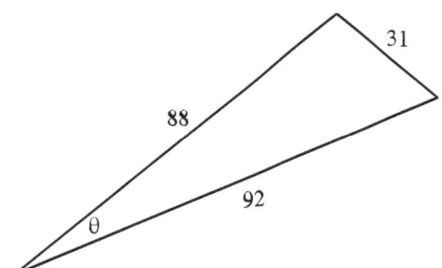

A. $31^2 = 88^2 + 92^2 - 2(88)(92)\cos\theta$

B. $92^2 = 88^2 + 31^2 - 2(88)(31)\cos\theta$

C. $88^2 = 31^2 + 92^2 - 2(31)(92)\cos\theta$

D. $\sin\theta = \dfrac{31}{92}$

E. $\cos\theta = \dfrac{88}{92}$

6. When $\dfrac{\cot\theta}{\cos\theta}$ is defined, it is equivalent to which of the following?

F. $\sin\theta$

G. $\dfrac{1}{\cos\theta}$

H. $\dfrac{1}{\sin\theta}$

J. $\dfrac{1}{\sin^2\theta}$

K. $\dfrac{\cos\theta}{\sin^2\theta}$

All-Angles Trigonometry
Advanced Problem Set 1

7. The sides of an acute triangle measure 11m, 17m and 20m. Which of the following equations when solved for θ gives the measure of the smallest angle of the triangle?
Note: for any triangle with sides of length a, b, and c that are opposite angles A, B and C, respectively, $\dfrac{\sin A}{a} = \dfrac{\sin B}{b} = \dfrac{\sin C}{c}$ and $c^2 = a^2 + b^2 - 2ab \cos C$.

A. $\dfrac{\sin \theta}{11} = \dfrac{1}{17}$

B. $\dfrac{\sin \theta}{11} = \dfrac{1}{20}$

C. $11^2 = 17^2 + 20^2 - 2(17)(20)\cos\theta$

D. $17^2 = 11^2 + 20^2 - 2(11)(20)\cos\theta$

E. $20^2 = 17^2 + 11^2 - 2(17)(11)\cos\theta$

8. In terms of θ, what is the area of the triangle in the figure below?

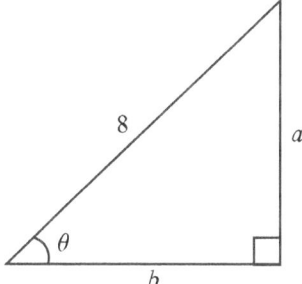

F. $4\sin\theta\cos\theta$

G. $8\sin\theta$

H. $32\sin\theta\cos\theta$

J. $\dfrac{1}{2}\tan\theta$

K. $64\sin\theta\cos\theta$

9. The sides of a triangle are 5, 7, and 9 inches long. What is the measure of the angle opposite the 7-inch-long side?

A. 50.70°

B. 85.54°

C. 94.46°

D. 95.74°

E. 129.30°

All-Angles Trigonometry
Advanced Problem Set 1

10. In $\triangle ABC$, $\dfrac{\sin A}{\sin B} = \dfrac{5}{12}$ and $\dfrac{\sin B}{\sin C} = \dfrac{3}{2}$. If angles A, B and C are opposite sides a, b, and c respectively, and the triangle has a perimeter of 5, then what is the length of a?

   F. 0.20
   G. 1.00
   H. 1.25
   J. 1.60
   K. 5.00

11. In the figure below, C is the center of the semicircle, and the area of the semicircle is 18π. What is the area of $\triangle ABC$ in terms of θ?

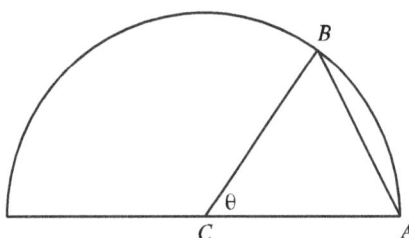

   A. $3\cos\theta$
   B. $3\sin\theta$
   C. $9\sin\theta$
   D. $6\sin\theta\tan\theta$
   E. $18\sin\theta$

All-Angles Trigonometry
Advanced Problem Set 1

12. A ball is thrown vertically into the air. The angles of elevation from points A and B, pictured below, are when the ball has reached a height of 35 feet. What is the distance between points A and B?

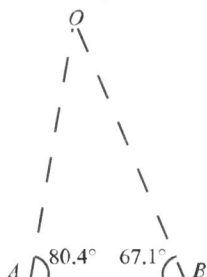

- **F.** 19.073
- **G.** 19.343
- **H.** 20.414
- **J.** 20.704
- **K.** 22.160

13. $\tan(\arcsin 0.6) =$
- **A.** 0.51
- **B.** 0.60
- **C.** 0.63
- **D.** 0.75
- **E.** 88.4

14. $\cos^{-1}(\sin 125°) =$
- **F.** $0°$
- **G.** $15°$
- **H.** $35°$
- **J.** $125°$
- **K.** $180°$

All-Angles Trigonometry
Advanced Problem Set 1

15. If $\cos(\arccos x) = \dfrac{\sqrt{3}}{5}$, then what is the value of x?

A. $\dfrac{\sqrt{3}}{5}$

B. $\dfrac{\sqrt{3}}{10}$

C. $\dfrac{3\sqrt{2}}{4}$

D. $\dfrac{5}{\sqrt{3}}$

E. $\dfrac{4}{3\sqrt{2}}$

16. In the figure below, $\dfrac{1}{\cos\theta} - \dfrac{1}{\sin\theta} =$

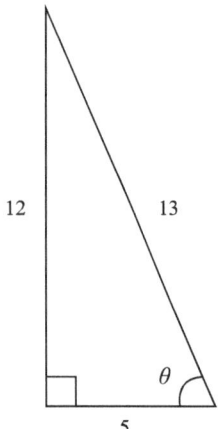

F. -0.5

G. 0.7

H. 1.5

J. 3.9

K. 43.72

17. If $\sin x = 0.812$, then $\csc x =$

A. 0.0142

B. 0.5837

C. 1.2315

D. 1.7133

E. 54.292

All-Angles Trigonometry
Advanced Problem Set 1

18. In the figure below, what is the degree measure of θ?

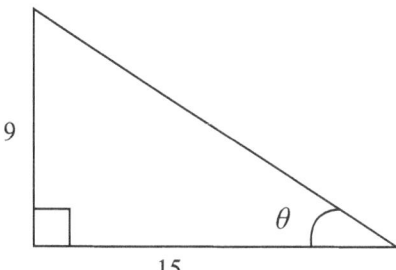

- **F.** 30.96°
- **G.** 36.87°
- **H.** 53.13°
- **J.** 59.04°
- **K.** 95.49°

19. In the figure below, if $\cot\theta = 1.4286$, then $\cos\theta = ?$

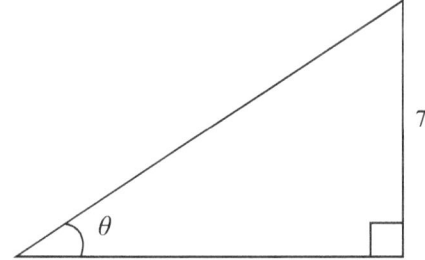

- **A.** 0.029
- **B.** 0.142
- **C.** 0.574
- **D.** 0.700
- **E.** 0.8192

All-Angles Trigonometry
Advanced Problem Set 1

20. In the figure below, $\dfrac{\sec B}{\tan A \sin B} =$

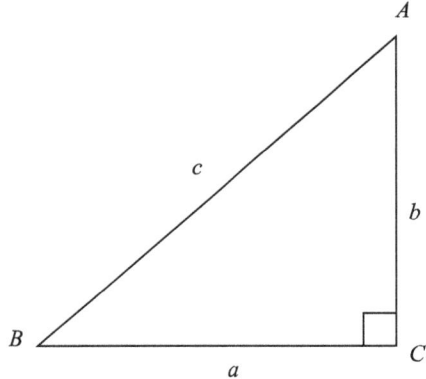

F. $\dfrac{c^2}{ab}$

G. $\dfrac{c^2}{a^2}$

H. 1

J. $\dfrac{b}{a}$

K. $\dfrac{ac^2}{b^3}$

21. In the figure below, what is the degree measure of θ?

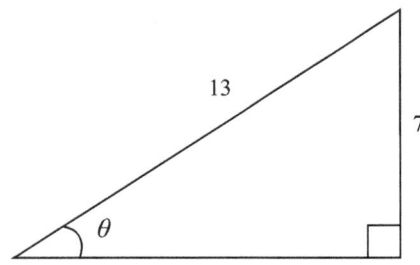

A. 0.53°

B. 1.86°

C. 28.30°

D. 32.58°

E. 57.42°

All-Angles Trigonometry
Advanced Problem Set 1

22. In the figure below, which of the folowing must be true?

    I. $\tan y = \dfrac{12}{5}$
    II. $\cot x = \cot y$
    III. $\sin x = \cos y$

    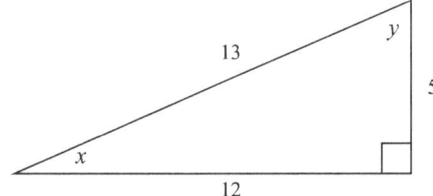

    F. I only
    G. III only
    H. I and III only
    J. I, II and III
    K. None of the above

23. In right $\triangle ABC$, $\angle B = 90°$, $\angle A = 56.1°$, and $CB = 12$. What is the length of the hypotenuse of $\triangle ABC$?

    A. 2.8
    B. 6.7
    C. 14.5
    D. 21.5
    E. 56

All-Angles Trigonometry
Advanced Problem Set 1

24. Building codes in Williamsburg require that wheelchair ramps must rise at an angle between $4.0°$ and $6.5°$ from the horizontal. If a wheelchair ramp rises exactly 4 feet as shown in the figure below, which of the following could be the length of the ramp?

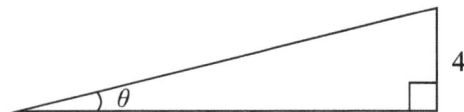

F. 12
G. 25
H. 34
J. 40
K. 62

25. In the figure below, $y =$

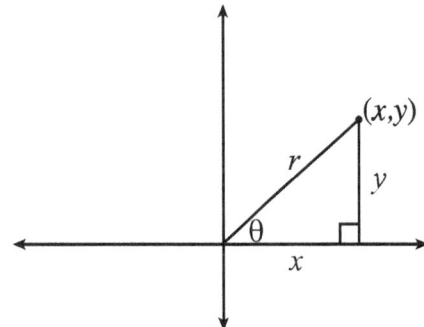

A. $r \tan \theta$
B. $r \cos \theta$
C. $r \sin \theta$
D. r^2
E. $x - r$

All-Angles Trigonometry
Advanced Problem Set 1

26. What is the range of $f(x) = -5\cos\frac{x}{4}$?

    F. All real numbers greater than or less than or equal to -5 and less than or equal to 0
    G. All real numbers less than or equal to -5 and greater than or equal to 5
    H. All real numbers greater than or equal to -5 and less than or equal to 5
    J. All real numbers greater than or equal to 0 and less than or equal to $\frac{1}{4}$
    K. All real numbers greater than or equal to 0 and less than or equal to 8π

27. If $f(x) = 2x + 3$ and $g(x) = f(\sin x) - f(\cos x)$, then $g(34°) =$

    A. -0.540
    B. 0.540
    C. 1.260
    D. 2.755
    E. 5.461

28. In the figure below, what is the value of θ?

    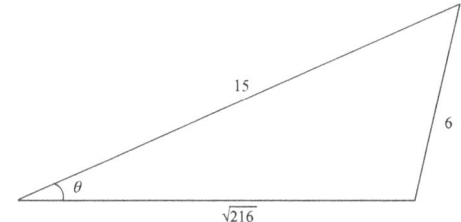

    F. 11.54°
    G. 22.21°
    H. 23.28°
    J. 75.52°
    K. 81.19°

All-Angles Trigonometry
Advanced Problem Set 1

Answer Key

#	Answer	Frequency	Difficulty
1	E	popular	1
2	J	popular	2
3	B	popular	2
4	H	popular	2
5	A	popular	1
6	H	popular	2
7	C	popular	2
8	H	popular	3
9	A	popular	3
10	G	popular	4
11	E	popular	4
12	J	popular	4
13	D	average	4
14	H	average	4
15	A	average	2
16	H	popular	2
17	C	popular	2
18	F	popular	2
19	E	popular	3
20	G	popular	4
21	D	popular	2
22	H	popular	2
23	C	popular	3
24	J	popular	4
25	C	popular	2
26	H	popular	3
27	A	average	4
28	H	popular	2

Converting Angles
Quick Drill

1. What is the reference angle for $120°$?

2. What is the reference angle for $338°$?

3. What is the reference angle for $\frac{3\pi}{4}$?

4. What is the reference angle for $\frac{7\pi}{6}$?

5. What is the reference angle for $-28°$?

6. Give the angle that is coterminal with $382°$ between $0°$ and $360°$.

7. Give the angle that is coterminal with $804°$ between $0°$ and $360°$.

8. Give the angle that is coterminal with $-108°$ between $0°$ and $360°$.

9. Give the angle that is coterminal with $\frac{13\pi}{2}$ between 0 and 2π.

10. Give the angle that is coterminal with $-\frac{2\pi}{3}$ between 0 and 2π.

11. Convert to radians: $30°$

12. Convert to radians: $228°$

13. Convert to degrees: $\frac{3\pi}{4}$

14. Convert to degrees: $\frac{6\pi}{5}$

15. Convert to degrees: $\frac{14\pi}{3}$

Converting Angles
Quick Drill

Answer Key

#	Answer
1	$60°$
2	$22°$
3	$\dfrac{\pi}{4}$
4	$\dfrac{\pi}{6}$
5	$28°$
6	$22°$
7	$84°$
8	$252°$
9	$\dfrac{\pi}{2}$
10	$\dfrac{4\pi}{3}$
11	$\dfrac{\pi}{6}$
12	$\dfrac{19\pi}{15}$
13	$135°$
14	$216°$
15	$840°$

Unit Circle
Quick Drill

1. $\sin\left(\dfrac{\pi}{6}\right) =$

2. $\cos(180°) =$

3. $\sin\left(-\dfrac{\pi}{2}\right) =$

4. $\cos\left(\dfrac{5\pi}{4}\right) =$

5. $\sin(120°) =$

6. Solve for θ such that $0 < \theta < 2\pi$
$\cos\theta = \dfrac{1}{2}$

7. Solve for θ such that $0 < \theta < 2\pi$
$\sin\theta = -1$

8. Solve for θ such that $0 < \theta < 2\pi$
$\cos\theta = -\dfrac{\sqrt{3}}{2}$

9. Solve for θ such that $0 < \theta < 2\pi$
$\sin\theta = -\dfrac{1}{2}$

10. Solve for θ such that $0 < \theta < 2\pi$
$\tan\theta = 1$

Unit Circle
Quick Drill

Answer Key

#	Answer
1	$\dfrac{1}{2}$
2	-1
3	-1
4	$-\dfrac{\sqrt{2}}{2}$
5	$\dfrac{\sqrt{3}}{2}$
6	$\theta = \dfrac{\pi}{3}, \dfrac{5\pi}{3}$
7	$\theta = \dfrac{3\pi}{2}$
8	$\theta = \dfrac{5\pi}{6}, \dfrac{7\pi}{6}$
9	$\theta = \dfrac{7\pi}{6}, \dfrac{11\pi}{6}$
10	$\theta = \dfrac{\pi}{4}, \dfrac{5\pi}{4}$

Unit Circle
Advanced Problem Set 2

1. If the value, to the nearest thousandth, of $\sin \theta$ is 0.906, which of the following could be true about θ?

   A. $\dfrac{\pi}{2} < \theta < \dfrac{5\pi}{6}$

   B. $\dfrac{5\pi}{6} < \theta < \dfrac{7\pi}{6}$

   C. $\dfrac{7\pi}{6} < \theta < \dfrac{9\pi}{6}$

   D. $\dfrac{8\pi}{6} < \theta < \dfrac{3\pi}{2}$

   E. $\dfrac{3\pi}{2} < \theta < \dfrac{7\pi}{4}$

2. What is the degree measure of the angle that has the radian measure $\dfrac{2\pi}{9}$?

   F. $20°$

   G. $40°$

   H. $80°$

   J. $\left(\dfrac{180 - 2\pi}{9}\right)°$

   K. $\left(\dfrac{360 - 2\pi}{9}\right)°$

3. If $\cos \theta = \dfrac{-5}{13}$ and $\dfrac{\pi}{2} < \theta < \pi$, then $\sin \theta = ?$

   A. $\dfrac{-5}{12}$

   B. $\dfrac{-12}{13}$

   C. $\dfrac{-13}{12}$

   D. $\dfrac{5}{12}$

   E. $\dfrac{12}{13}$

Unit Circle
Advanced Problem Set 2

4. Given that $\sin A = \frac{5}{13}$ and $0° \leq A < 360°$, what are all possible values of $\cos A$?

   F. $-\frac{5}{13}$

   G. $-\frac{5}{13}$ and $\frac{5}{13}$

   H. $\frac{12}{13}$

   J. $-\frac{12}{13}$

   K. $-\frac{12}{13}$ and $\frac{12}{13}$

5. If $0 \leq \theta < 2\pi$ and $\sin^2 \theta + \sin \theta = \frac{3}{4}$, what is $\cos \theta$?

   A. $\frac{\sqrt{3}}{2}$

   B. $\sqrt{2}$

   C. $\frac{\pi}{6}$

   D. $\frac{1}{2}$

   E. 0

6. If $0° \leq x \leq 90°$, and $2\sin^2 x - 1 = 0$, then $x =$

   F. $0°$

   G. $30°$

   H. $45°$

   J. $60°$

   K. $90°$

7. If $\sin x = 3 \cos x$, then what is the radian value of x?

   A. 0.33

   B. 0.52

   C. 1.25

   D. ± 0.33

   E. ± 1.25

Unit Circle
Advanced Problem Set 2

8. If $5\tan x + 4 = 0$ on $\pi \leq x \leq \dfrac{3\pi}{2}$, then $x =$

   F. -0.800
   G. -0.675
   H. -0.068
   J. 0.608
   K. 5.608

9. If $x \geq 0$ and $\cos x = \sin(2x)$, then x could equal

   A. $0°$
   B. $15°$
   C. $30°$
   D. $45°$
   E. $60°$

10. If $\cot\theta = \dfrac{3}{4}$, then $\sin\theta = ?$

    F. $\dfrac{3}{5}$
    G. $\pm\dfrac{3}{5}$
    H. $\dfrac{4}{5}$
    J. $\pm\dfrac{4}{5}$
    K. $\dfrac{5}{4}$

11. If $\pi \leq \theta \leq \dfrac{3\pi}{2}$ and $\tan\theta = 0.839$, then $\sec\theta = ?$

    A. -1.305
    B. -0.766
    C. 0.766
    D. 1.305
    E. 4.447

DO YOUR FIGURING HERE

Unit Circle
Advanced Problem Set 2

12. $\sin(90°) =$

   F. $\cos(90°)$

   G. $2\sin(45°)\cos(45°)$

   H. $\sin(45°)\cos(45°)$

   J. $\sin^2(45°)$

   K. $\tan(90°)$

13. What does an angle of $-\dfrac{3\pi}{8}$ radians equal in degree measure?

   A. $-67.5°$

   B. $-0.021°$

   C. $-0.375°$

   D. $0.375°$

   E. $67.5°$

14. If $\cos x = -\dfrac{4}{5}$ and $\dfrac{\pi}{2} < x < \pi$, what is the value of $\cos 2x = ?$

   F. $-\dfrac{8}{5}$

   G. $-\dfrac{2}{5}$

   H. $-\dfrac{7}{25}$

   J. $\dfrac{7}{25}$

   K. $\dfrac{8}{5}$

15. If $\sin x = \dfrac{3}{5}$ and $0 < x < \pi$, then $\tan 2x =$

   A. -3.43

   B. 0.75

   C. 2.21

   D. 3.43

   E. 4.43

Unit Circle
Advanced Problem Set 2

16. What is the value of x if $\pi \leq x \leq \frac{3\pi}{2}$ and $\sin x = 8 \cos x$?

- F. -6.800
- G. -0.124
- H. 1.446
- J. 4.588
- K. 6.159

17. If $0 < x < \frac{\pi}{2}$ and $\sin x = 0.590$, what is the value of $\cos \frac{x}{2}$?

- A. 0.295
- B. 0.316
- C. 0.892
- D. 0.951
- E. 1.250

18. If $\sin \theta = t$, then for all θ in the interval $\frac{\pi}{2} < \theta < \pi$, $\tan \theta =$

- F. $\sqrt{1-t^2}$
- G. $\dfrac{1-t^2}{t}$
- H. $\dfrac{\sqrt{1-t^2}}{t}$
- J. $\dfrac{t}{\sqrt{1-t^2}}$
- K. $-\dfrac{t}{\sqrt{1-t^2}}$

Unit Circle
Advanced Problem Set 2

19. If $\cos\theta = \dfrac{5}{13}$ and $\dfrac{3\pi}{2} < \theta < 2\pi$, then $\tan\theta =$

   A. $-\dfrac{12}{5}$

   B. $-\dfrac{5}{12}$

   C. $-\dfrac{5}{13}$

   D. $\dfrac{5}{12}$

   E. $\dfrac{12}{5}$

20. If $\pi \leq x \leq \dfrac{3\pi}{2}$ and $\cot x = \dfrac{3}{a}$, then $\sec x =$

   F. $\dfrac{3}{a^2+9}$

   G. $\dfrac{a}{a^2+9}$

   H. $\dfrac{\sqrt{a^2+9}}{3}$

   J. $\dfrac{\sqrt{a^2+9}}{a}$

   K. $-\dfrac{\sqrt{a^2+9}}{3}$

21. What does the angle $585°$ equal in radian measure?

   A. $\dfrac{3\pi}{4}$

   B. $\dfrac{5\pi}{4}$

   C. $\dfrac{13\pi}{8}$

   D. $\dfrac{13\pi}{4}$

   E. $\dfrac{13\pi}{2}$

DO YOUR FIGURING HERE

Unit Circle
Advanced Problem Set 2

22. If $\cos\theta = -\dfrac{3}{5}$ and $\dfrac{\pi}{2} \leq \theta \leq \pi$, then $\sin(2\theta) =$

F. $-\dfrac{24}{25}$

G. $-\dfrac{4}{5}$

H. $\dfrac{4}{5}$

J. $\dfrac{24}{25}$

K. $\dfrac{8}{5}$

23. If $\dfrac{\pi}{2} \leq \theta \leq \pi$ and $\csc\theta = 3$, then $\tan\theta =$

A. $-\sqrt{8}$

B. $-\dfrac{1}{\sqrt{8}}$

C. $\dfrac{1}{8}$

D. $\dfrac{1}{\sqrt{8}}$

E. $\sqrt{8}$

24. For an angle with measure β in a right triangle, $\sec\beta = \dfrac{37}{35}$ and $\cot\beta = \dfrac{35}{12}$. What is $\csc\beta$?

F. $\dfrac{12}{35}$

G. $\dfrac{12}{37}$

H. $\dfrac{35}{37}$

J. $\dfrac{37}{12}$

K. $\dfrac{37}{5}$

Unit Circle
Advanced Problem Set 2

25. If $0 \leq x \leq \pi$ and $\cos x < 0$, which of the following must be true?

$$\text{I. } \sin x < 0$$
$$\text{II. } \sec x < 0$$
$$\text{III. } \sin x + \cos x > 0$$

- **A.** II only
- **B.** III only
- **C.** I and III only
- **D.** II and III only
- **E.** None of the above

26. If $\pi < \theta < \dfrac{3\pi}{2}$ and $\sin \theta = -\dfrac{12}{13}$, then $\tan \theta =$

- **F.** $-\dfrac{5}{12}$
- **G.** $\dfrac{5}{12}$
- **H.** $\dfrac{12}{5}$
- **J.** $-\dfrac{12}{5}$
- **K.** $\dfrac{12}{13}$

27. If $180° < \alpha < 360°$ and the $\cos \alpha = -\dfrac{4}{5}$, then $\sin \alpha =$

- **A.** $-\dfrac{3}{5}$
- **B.** $-\dfrac{4}{5}$
- **C.** $\dfrac{3}{5}$
- **D.** $\dfrac{4}{5}$
- **E.** $-\dfrac{5}{3}$

DO YOUR FIGURING HERE

Unit Circle
Advanced Problem Set 2

28. If $\cos\theta = \dfrac{3}{7}$ and θ is the measure of a positive, acute angle then $\tan\theta =$

F. $\dfrac{7}{3}$

G. $\dfrac{3}{2\sqrt{10}}$

H. $\dfrac{3}{10}$

J. $\dfrac{2\sqrt{10}}{7}$

K. $\dfrac{2\sqrt{10}}{3}$

29. Given that $\cos C = \dfrac{12}{13}$ and $0° \leq C \leq 360°$, what are all possible values of $\sin C$?

A. $\dfrac{5}{12}$ only

B. $-\dfrac{12}{13}$ and $\dfrac{12}{13}$

C. $\dfrac{5}{13}$ and $-\dfrac{5}{13}$

D. $\dfrac{13}{5}$

E. $-\dfrac{13}{5}$ and $\dfrac{13}{5}$

30. If $\sec\theta = \dfrac{2}{5t}$, then where defined, $\sin\theta =$

F. $\pm\dfrac{5t}{2}$

G. $\pm\sqrt{2-5t}$

H. $\pm\sqrt{4-25t^2}$

J. $\pm\dfrac{\sqrt{2-5t}}{2}$

K. $\pm\dfrac{\sqrt{4-25t^2}}{2}$

Unit Circle
Advanced Problem Set 2

31. What are the (x, y) coordinates of point A in the unit circle below?

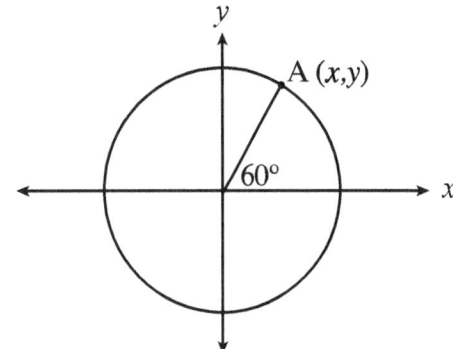

A. $\left(\dfrac{\sqrt{3}}{2}, \dfrac{\sqrt{3}}{2}\right)$

B. $\left(\dfrac{1}{2}, \dfrac{1}{2}\right)$

C. $\left(\dfrac{\sqrt{3}}{2}, \dfrac{1}{2}\right)$

D. $\left(\dfrac{1}{2}, \dfrac{\sqrt{2}}{2}\right)$

E. $\left(\dfrac{1}{2}, \dfrac{\sqrt{3}}{2}\right)$

DO YOUR FIGURING HERE

Unit Circle
Advanced Problem Set 2

Answer Key

#	Answer	Frequency	Difficulty
1	A	popular	2
2	G	popular	2
3	E	popular	2
4	K	popular	2
5	A	popular	3
6	H	popular	3
7	C	average	4
8	G	average	4
9	C	average	4
10	J	popular	3
11	A	popular	4
12	G	average	4
13	A	popular	2
14	J	popular	3
15	D	popular	3
16	H	popular	3
17	D	popular	2
18	K	popular	3
19	A	popular	3
20	K	popular	3
21	D	popular	1
22	F	popular	3
23	B	popular	3
24	J	popular	4
25	A	popular	4
26	H	popular	3
27	A	popular	3
28	K	popular	3
29	C	popular	3
30	K	popular	2
31	E	popular	2

Graphing Trig Functions
Quick Drill

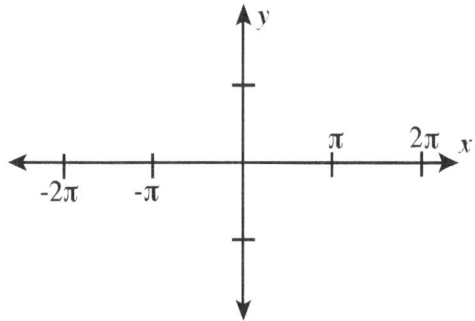

1. Graph $\sin x$.

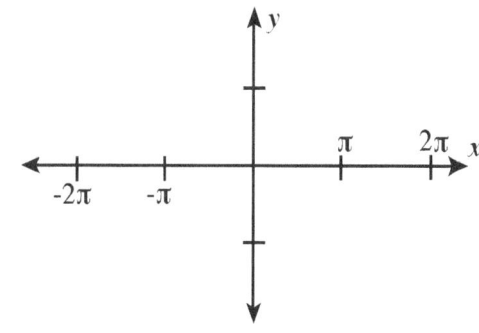

3. Graph $\tan x$.

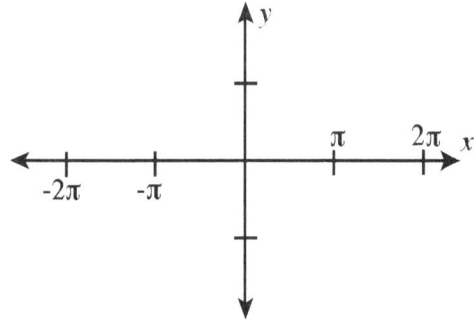

2. Graph $\cos x$.

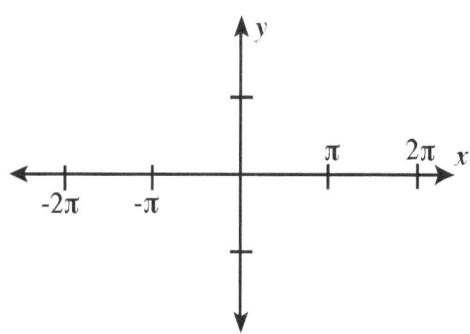

4. Graph $3\sin x$.

Graphing Trig Functions
Quick Drill

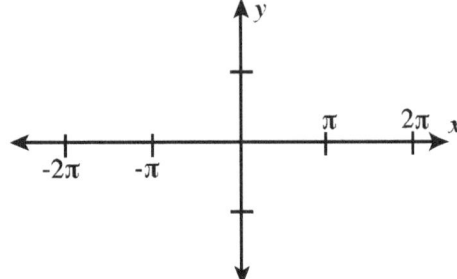

5. Graph $3\cos 2x$.

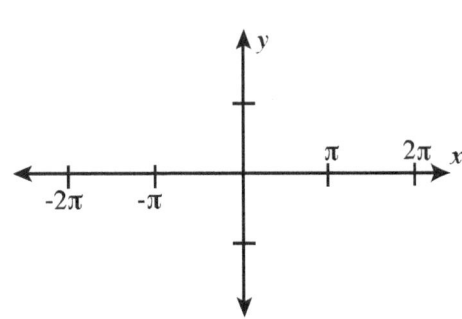

6. Graph $\sin \frac{1}{2}x$.

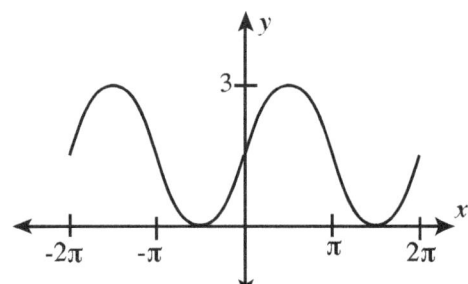

7. Write an equation for the graph.

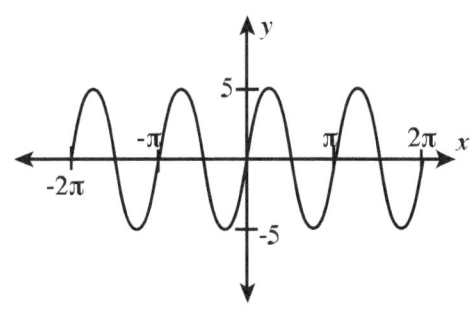

8. Write an equation for the graph.

9. Write an equation for the graph.

Graphing Trig Functions
Quick Drill

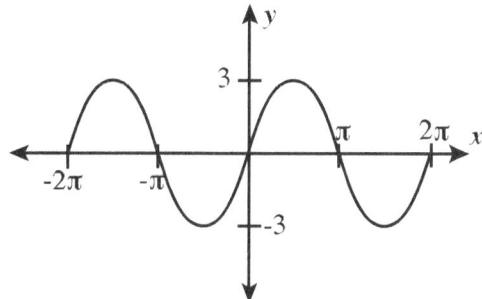

10. Write an equation for the graph using cosine.

Graphing Trig Functions
Quick Drill

Answer Key

#	Answer
1	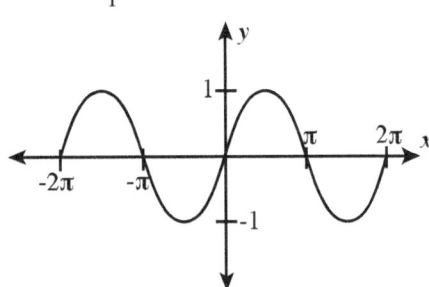
2	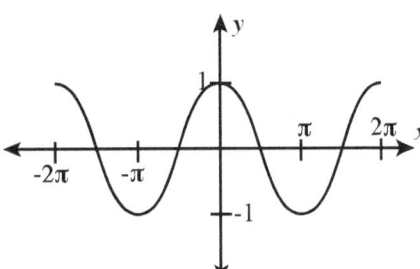
3	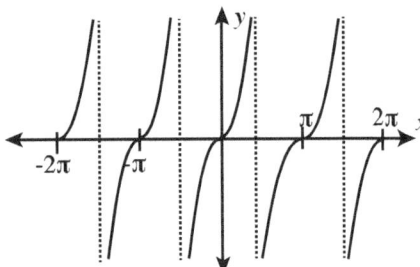
4	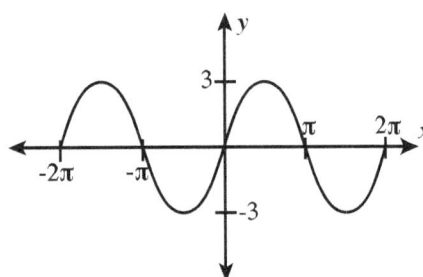
5	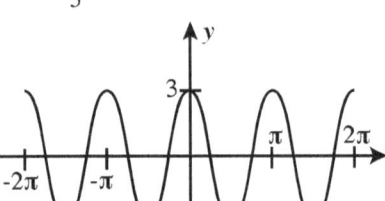
6	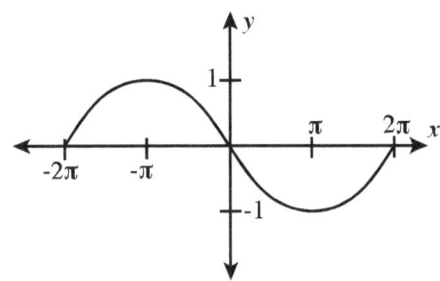
7	$y = 2\cos x$
8	$y = 1.5 \sin x + 1.5$
9	$y = 5 \sin 2x$
10	$y = 3 \cos\left(x - \dfrac{\pi}{2}\right)$ or $y = 3\cos\left(x + \dfrac{3\pi}{2}\right)$

Trig Identities
Quick Drill

1. Express in terms of sine and cosine.
   $\sec x$

2. Express in terms of sine and cosine.
   $\tan x$

3. Express in terms of sine and cosine.
   $\csc x$

4. Express in terms of sine and cosine.
   $\cot x$

5. $\sin^2 x + \cos^2 x =$

6. Express in terms of sine and cosine.
   $\cos x \tan x$

7. Express in terms of sine and cosine.
   $\sin x \sec x \cot x$

8. Express in terms of sine and cosine.
   $\dfrac{\tan x}{\sec x}$

9. Express in terms of sine and cosine.
   $\dfrac{\sec x}{\csc x}$

10. Express in terms of sine and cosine.
    $\dfrac{\sec^2 x \sin x}{\tan x}$

11. Express in terms of sine and cosine.
    $\dfrac{1 - \sin^2 x}{\cos^2 x} + \dfrac{1 - \cos^2 x}{\sin^2 x}$

Trig Identities
Quick Drill

13. Solve for x where $0 < x \leq 2\pi$.
    $\sin^2 x = 4 - 5\cos^2 x$

14. Solve for x where $0 < x \leq 2\pi$.
    $3\tan^2 x + 2 = 3$

15. Solve for x where $0 < x \leq 2\pi$.
    $2\sin^2 x + 3\cos x - 3 = 0$

16. $\sqrt{\dfrac{9 - 9\sin^2 x}{\cos^2 x} + \dfrac{16 - 16\cos^2 x}{\sin^2 x}} =$

Trig Identities
Quick Drill

Answer Key

#	Answer
1	$\dfrac{1}{\cos x}$
2	$\dfrac{\sin x}{\cos x}$
3	$\dfrac{1}{\sin x}$
4	$\dfrac{\cos x}{\sin x}$ or $\dfrac{1}{\tan x}$
5	1
6	$\sin x$
7	1
8	$\sin x$
9	$\dfrac{\sin x}{\cos x}$ or $\tan x$
10	$\dfrac{1}{\cos x}$ or $\sec x$
11	2
12	$\pi, \dfrac{3\pi}{2}, 2\pi$
13	$\dfrac{\pi}{6}, \dfrac{5\pi}{6}, \dfrac{7\pi}{6}, \dfrac{11\pi}{6}$
14	$\dfrac{\pi}{6}, \dfrac{5\pi}{6}, \dfrac{7\pi}{6}, \dfrac{11\pi}{6}$
15	$\dfrac{\pi}{3}, \dfrac{5\pi}{3}, 2\pi$
16	5

Trigonometry Graphing and IDs
Advanced Problem Set 3

1. Which of the following correctly describes the graph of $y = 2\sin(3x)$?

    A. It is even

    B. It is odd

    C. It exhibits symmetry over the x-axis

    D. It has a maximum of 3

    E. It has a frequency of 2

2. Which of the following functions has the greatest frequency?

    F. $y = \sin x$

    G. $y = \tan(2x)$

    H. $y = \cos(3x)$

    J. $y = 9\cos x$

    K. $y = \cos(x + 5)$

3. What is the smallest possible period of the function $y = 2 + 3\cos(2x + 12)$?

    A. 2

    B. π

    C. 2π

    D. 12

    E. 4π

4. Which of the following functions has the greatest period?

    F. $y = \sin\left(\dfrac{x}{2}\right)$

    G. $y = \tan\left(\dfrac{x}{2}\right)$

    H. $y = \cos(3x)$

    J. $y = 9\cos x$

    K. $y = \cos(x + 5)$

Trigonometry Graphing and IDs
Advanced Problem Set 3

5. Which of the following has an amplitude of 5?

   A. $\dfrac{1}{5\sin x}$

   B. $2\sin x$

   C. $\sin\left(\dfrac{2\pi}{5x}\right)$

   D. $5\cos x$

   E. $\dfrac{5}{(2\pi)\cos x}$

6. What is the period of $y = 4\pi \tan(\pi(x+2))$?

   F. 1

   G. 2

   H. π

   J. 2π

   K. 4π

7. Which of the following statements is true about the graph of $y = a\sin(fx + b)$?

   A. As f increases, the period decreases.

   B. As f decreases, the period decreases.

   C. As a increases, the period increases.

   D. As b increases, the period increases.

   E. As f increases, the amplitude increases.

8. Which of the following functions has the greatest amplitude?

   F. $y = \sin(9x)$

   G. $y = \left(\dfrac{1}{9}\right)\sin x$

   H. $y = \left(\dfrac{9}{2}\right)\sin(9x)$

   J. $y = \left(\dfrac{9}{2}\right)\sin(x+5)$

   K. $y = 9\sin x$

Trigonometry Graphing and IDs
Advanced Problem Set 3

9. Which of the following trigonometric functions has an amplitude of 3?

   A. $f(x) = 3\sin x + 2$

   B. $f(x) = 3\cot x$

   C. $f(x) = \sin(x+3)$

   D. $f(x) = \sin(3x)$

   E. $f(x) = \tan x$

10. Which of the following is an even function?

    I. $y = \sin x \cos x$
    II. $y = \cos x$
    III. $y = |\tan x|$

    F. I only

    G. II only

    H. III only

    J. I and III only

    K. II and III only

11. Which of the following is equal to the amplitude of $y = 4\sin x$?

    A. $4\sin\pi$

    B. $4\sin\left(\dfrac{\pi}{2}\right)$

    C. $4\sin\left(\dfrac{\pi}{2}\right) - 4\sin\left(\dfrac{3\pi}{2}\right)$

    D. $4\sin(2\pi)$

    E. $4\sin(2\pi) - 4\sin 0$

12. What is the period of $y = \tan x$?

    F. 1

    G. 2

    H. π

    J. 2π

    K. $\tan x$

Trigonometry Graphing and IDs
Advanced Problem Set 3

DO YOUR FIGURING HERE

13. How many times does the graph of $y = \sin x$ complete a full cycle from 0 to 32π on the x-axis?

A. 9

B. 16

C. 32

D. 18π

E. 32π

14. What is the value of $y = \cos x$ when its graph is three quarters of the way through its first period starting at 0?

F. -2

G. $\frac{\sqrt{2}}{2}$

H. $\frac{\sqrt{3}}{2}$

J. 0

K. 1

15. Which of the following graphs has a period of π?

A. $y = \sin x$

B. $y = \cos x$

C. $y = \tan x$

D. $y = \frac{1}{\sin x}$

E. $y = \pi \sin x$

16. In the equation $y = z \sin(2cx)$, what is the frequency?

F. 2

G. c

H. z

J. $\frac{c}{z}$

K. $2c$

Trigonometry Graphing and IDs
Advanced Problem Set 3

17. Which of the following functions is symmetrical with respect to the origin?

    A. $y = \sin\left(x + \dfrac{\pi}{4}\right)$

    B. $y = \cos x$

    C. $y = \pi$

    D. $y = (\sin x)^2$

    E. $y = \tan x$

18. If $f(x) = \sin\left(x^2 + 2x + \dfrac{e}{9}\right)$, what is the maximum value of the function $f(x)$?

    F. 1

    G. 2

    H. π

    J. 2π

    K. $\sin\left(\dfrac{e}{9}\right)$

19. If $15\sin^2 x + 2\sin x - 1 = 0$ over the interval $180° \leq x \leq 360°$, then $x =$

    A. -19.5° or 340.5°

    B. -19.5° or 11.5°

    C. 340.5° or 250.5°

    D. 199.5° or 340.5°

    E. 199.5° or 191.5°

20. If $24\cos^2 \theta + 8\cos\theta - 2 = 0$, then what is the smallest positive value of θ?

    F. 20.4°

    G. 60°

    H. 80.4°

    J. 90.6°

    K. 120°

Trigonometry Graphing and IDs
Advanced Problem Set 3

21. The functions $y = \sin x$ and $y = \sin(x + a) + b$, for constants a and b, are graphed in the standard xy-coordinate plane below. The functions have the same maximum value. Which of the following statements about the values of a and b could be true?

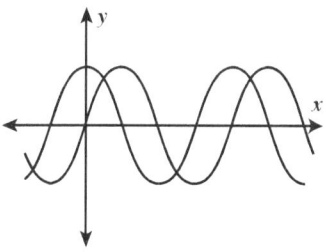

A. $a < 0$ and $b = 0$

B. $a < 0$ and $b > 0$

C. $a = 0$ and $b > 0$

D. $a > 0$ and $b < 0$

E. $a > 0$ and $b > 0$

Trigonometry Graphing and IDs
Advanced Problem Set 3

Answer Key

#	Answer	Frequency	Difficulty
1	B	average	2
2	H	average	2
3	B	average	2
4	F	average	2
5	D	average	2
6	F	average	2
7	A	average	2
8	K	average	2
9	A	popular	2
10	K	average	2
11	B	average	2
12	H	average	2
13	B	average	2
14	J	average	2
15	C	average	2
16	K	average	2
17	E	average	2
18	F	average	2
19	D	average	4
20	H	average	4
21	A	popular	1

Polynomial Algebra
Advanced Problem Set 4

1. What is the sum of the 4 binomials listed below?
   $3x^2 - 2x, 2x^2 + 5, 5x - 16, 7x^2 + 8x$

   A. $12x^2 - 11x + 11$

   B. $12x^2 + 11x + 11$

   C. $12x^2 + 11x - 11$

   D. $12x^2 - 11x - 11$

   E. $-12x^2 + 11x - 11$

2. For a certain quadratic equation, $ax^2 + bx + c = 0$, the 2 solutions are $x = \frac{2}{9}$ and $x = \frac{-4}{5}$. Which of the following could be factors of $ax^2 + bx + c$?

   F. $(9x + 2)(5x - 4)$

   G. $(-9x - 2)(-4x - 5)$

   H. $(2x - 9)(4x - 5)$

   J. $(-9x + 2)(5x + 4)$

   K. $(-2x + 9)(-5x + 4)$

3. What is the sum of the 2 solutions to the equation $12x^2 - x - 63 = 0$?

   A. $\frac{-1}{12}$

   B. $\frac{-2}{7}$

   C. $\frac{1}{12}$

   D. $\frac{2}{7}$

   E. $\frac{55}{12}$

DO YOUR FIGURING HERE

Polynomial Algebra
Advanced Problem Set 4

4. For all x and y, $(3x + 2y)(4x^2 - 3y) = ?$
   - F. $12x^3 - 6y^2$
   - G. $12x^3 + 8x^2y - 6y^2$
   - H. $12x^3 - 8x^2y + 6y^2$
   - J. $12x^3 + 9xy - 8x^2y - 6y^2$
   - K. $12x^3 - 9xy + 8x^2y - 6y^2$

5. What are the real solutions to the equation $|x|^2 - 5|x| + 6 = 0$?
   - A. ± 2
   - B. ± 3
   - C. 2 and 3
   - D. -2 and -3
   - E. ± 2 and ± 3

6. For all x, $(4x + 6)^2 = ?$
   - F. $16x + 12$
   - G. $16x^2 + 12$
   - H. $16x^2 + 36$
   - J. $16x^2 + 24x + 36$
   - K. $16x^2 + 48x + 36$

7. Which of the following is a factored form of $6x^2 + x - 15$?
   - A. $(3x - 5)(2x + 3)$
   - B. $(3x + 5)(2x - 3)$
   - C. $(2x - 5)(3x + 3)$
   - D. $(2x + 5)(3x - 3)$
   - E. $(2x - 5)(3x - 3)$

DO YOUR FIGURING HERE

Polynomial Algebra
Advanced Problem Set 4

8. For all x, $(4x+9)^2 = ?$
   - F. $8x + 18$
   - G. $8x^2 + 18$
   - H. $16x^2 + 81$
   - J. $16x^2 + 36x + 81$
   - K. $16x^2 + 72x + 81$

9. If $x = -5$ is one solution to the equation $x^2 + ax + 45 = 0$, then the other solution is?
   - A. -9
   - B. -3
   - C. -5
   - D. 3
   - E. 9

10. For what integer k are both solutions of the equation $x^2 + kx + 23 = 0$ positive integers?
    - F. -24
    - G. -22
    - H. 1
    - J. 22
    - K. 24

11. For all $a \neq 0$, $\dfrac{(4a^3)^2 + 2a^5 - 6(a^3)^4}{2a^2} = ?$
    - A. $4a^3 + 2a^{\frac{5}{2}} - 6a^6$
    - B. $4a^8 + 2a + 6a^3$
    - C. $8a^3 + a^{\frac{5}{2}} + 3a^6$
    - D. $8a^4 + a^3 - 3a^{10}$
    - E. $8a^8 + a + 6a^3$

DO YOUR FIGURING HERE

Polynomial Algebra
Advanced Problem Set 4

12. The sum of $(2x^4 + 3x^2 - 2x + 5)$ and which polynomial results in the polynomial $2x^4 + x^2 + 2x + 12$?

  F. $-4x^4 + 2x^2 + 6$

  G. $-4x^4 - 2x^2 + 7$

  H. $4x^4 - 2x^2 + 4x + 7$

  J. $2x^2 + 4x + 6$

  K. $-2x^2 + 4x + 7$

13. For all $x^2 \neq 144$, $\dfrac{(x-12)^2}{x^2 - 144} = ?$

  A. $\dfrac{x-12}{x+12}$

  B. $\dfrac{1}{x-12}$

  C. $\dfrac{1}{x+12}$

  D. $\dfrac{-1}{12}$

  E. $\dfrac{1}{12}$

14. Which of the following expressions is a factor of the polynomial $x^2 + 7x - 30$?

  F. $x - 4$

  G. $x + 5$

  H. $x - 6$

  J. $x + 10$

  K. $x + 3$

15. If 3 is a solution to the equation $x^2 + hx + 15 = 0$, what does h equal?

  A. -8

  B. -5

  C. -3

  D. 5

  E. 8

DO YOUR FIGURING HERE

Polynomial Algebra
Advanced Problem Set 4

16. Which of the following expressions is equivalent to $5x^4 + 30x^3 - 35x^2$?

    F. $(x-3)(x-2)$

    G. $(x-7)(x-1)$

    H. $(x+7)(x-1)$

    J. $5x(x+7)(x-1)$

    K. $5x^2(x+7)(x-1)$

17. What quantity must be added to $5x^2 + 2x - 7$ to obtain $x^3 - 17$?

    A. $x^3 - 10$

    B. $-x^3 + 10$

    C. $x^3 + 5x^2 + 2x - 10$

    D. $x^3 - 5x^2 - 2x - 10$

    E. $-x^3 - 5x^2 + 2x + 10$

18. $(3x+7)(4x-8)$ is equivalent to:

    F. $7x^2 - 1$

    G. $7x^2 - 56$

    H. $7x^2 - 56x - 1$

    J. $12x^2 + 4x - 56$

    K. $12x^2 - 56$

19. Which of the following is a factor of $x^2 - x - 42$?

    A. $x - 7$

    B. $x - 6$

    C. $x - 3$

    D. $x + 3$

    E. $x + 7$

DO YOUR FIGURING HERE

Polynomial Algebra
Advanced Problem Set 4

20. What is the sum of the polynomials $4a^3b + 3ab^2$ and $-a^2b + ab^2$?

   F. $4a^3b - a^2b + 3ab^2$

   G. $4a^3b - a^2b + 4ab^2$

   H. $3a^3b + 4ab^2$

   J. $3a^3b + 3ab^2$

   K. $-4a^3b + 2ab^2$

21. If $(x+k)^2 = x^2 + 14x + k^2$ for all real numbers x, then $k = ?$

   A. -7

   B. 7

   C. 14

   D. 28

   E. 35

22. Which of the following is a value of t for which $(t-5)(t+3) = 0$?

   F. 2

   G. 3

   H. 4

   J. 5

   K. 6

23. For all $x > 2$, $\dfrac{4x - 2x^2}{-x^2 - x + 6} = ?$

   A. $\dfrac{1}{x+3}$

   B. $\dfrac{x}{x+3}$

   C. $\dfrac{2x}{-x^2 - x + 6}$

   D. $\dfrac{2x}{-x+2}$

   E. $\dfrac{2x}{x+3}$

Polynomial Algebra
Advanced Problem Set 4

24. Which of the following is not a solution of $(x-2)(x+4)(x+7)(x-5) = 0$?

   F. -7
   G. -5
   H. -4
   J. 2
   K. 5

25. What is the value of c if $x + 3$ is a factor of $x^3 + 5x^2 + 3cx + 9$?

   A. -3
   B. -2
   C. 1
   D. 2
   E. 3

26. $(7a + 6b + 5c) - (8a + 3b - 3c)$ is equivalent to:

   F. $15a + 9b + 2c$
   G. $15a + 9b + 8c$
   H. $-a + 3b + 2c$
   J. $-a + 3b + 8c$
   K. $-a - 3b - 8c$

27. The expression $(8c - 3d)(c + 5d)$ is equivalent to:

   A. $9c^2 - 8d^2$
   B. $9c^2 - 3dc - 8d^2$
   C. $8c^2 - 3dc + 15d^2$
   D. $8c^2 + 37dc + 15d^2$
   E. $8c^2 + 37dc - 15d^2$

Polynomial Algebra
Advanced Problem Set 4

28. $t^2 - 45t + 32 - 63t^2 + 59t$ is equivalent to

   F. $-16t^2$

   G. $-16t^6$

   H. $-62t^2 - 14t + 32$

   J. $-62t^2 + 14t + 32$

   K. $-63t^2 + 2t + 32$

29. The expression $(5z - 3)(z - 7)$ is equvalent to:

   A. $5z^2 - 10$

   B. $5z^2 + 21$

   C. $5z^2 - 3z - 10$

   D. $5z^2 + 38z + 21$

   E. $5z^2 - 38z + 21$

30. $(2x^3 - 5x^2 + 3x) - (x^2 + 4x - 7)$ simplifies to:

   F. $2x^3 - 4x^2 + 7x + 7$

   G. $2x^3 - 4x^2 - x - 7$

   H. $2x^3 - 6x^2 + 7x - 7$

   J. $2x^3 - 6x^2 - x + 7$

   K. $2x^3 - 6x^2 - x - 7$

31. Which of the following expressions would represent the area of a right triangle if the base was $(x + 3)$ and the height was $(x - 2)$?

   A. $2x + 1$

   B. $\dfrac{2x + 1}{2}$

   C. $x^2 - 6$

   D. $\dfrac{(x^2 - x - 6)}{2}$

   E. $\dfrac{(x^2 + x - 6)}{2}$

Polynomial Algebra
Advanced Problem Set 4

32. In the equation $x^2 + mx + n = 0$, m and n are integers. The only possible value for x is -5. What is the value of m?

    F. -10

    G. -5

    H. 5

    J. 10

    K. 25

33. What polynomial must be added to $x^2 - 3x + 8$ so the sum is $5x^2 + 9x$?

    A. $6x^2 + 6x + 8$

    B. $5x^2 + 12x + 8$

    C. $5x^2 + 12x - 8$

    D. $4x^2 + 12x - 8$

    E. $4x^2 + 6x - 8$

34. For $x^2 \neq 36$, $\dfrac{(x-6)^2}{x^2 - 36} = ?$

    F. $\dfrac{-1}{6}$

    G. $\dfrac{1}{6}$

    H. $\dfrac{1}{x+6}$

    J. $\dfrac{1}{x-6}$

    K. $\dfrac{x-6}{x+6}$

DO YOUR FIGURING HERE

Polynomial Algebra
Advanced Problem Set 4

35. For $x \neq 6$ and $x \neq -6$, $\dfrac{5x}{x^2 - 36} + \dfrac{5x}{6 - x}$ is equivalent to:

   A. $\dfrac{-5x^2}{x^2 - 36}$

   B. $\dfrac{-5x^2 - 25x}{x^2 - 36}$

   C. $\dfrac{25x^2 - 30x}{x^2 - 36}$

   D. $\dfrac{-30x}{x^2 - 36}$

   E. $\dfrac{-5x + 36}{x^2 - 36}$

36. For all x, $(x^2 + 5x + 6)(x - 3) = ?$

   F. $x^3 + 2x^2 - 9x - 18$

   G. $x^3 + 2x^2 + 21x + 18$

   H. $x^3 + 5x^2 + 9x - 18$

   J. $x^3 + 5x^2 + 9x + 18$

   K. $x^3 + 2x - 18$

37. Which of the following is equivalent to $(4x + 7)(5x - 3)$?

   A. $20x^2 - 21$

   B. $20x^2 + 23x - 21$

   C. $20x^2 - 23x - 21$

   D. $20x^2 + 47x - 21$

   E. $20x^2 + 47x + 21$

38. The expression given below is equivalent to which of the following expressions?
   $(2x^3 + x^2 - 7) - 3(x^4 - 5x^3 + x^2 + 6x + 4)$

   F. $x^{15} - 3$

   G. $-3x^4 + 17x^3 - 2x^2 - 18x - 19$

   H. $-3x^4 - 3x^3 + 2x^2 - 6x - 3$

   J. $-3x^4 - 13x^3 + 4x^2 + 18x + 5$

   K. $-6x^{15}$

Polynomial Algebra
Advanced Problem Set 4

39. $(x^3 + 27)(x^3 - 27)$ is equivalent to:

    A. $2x^3$

    B. $-x^6$

    C. $x^6 - 729$

    D. $x^9 - 729$

    E. $x^6 - 54x^3 - 729$

40. What is the coefficient of x^9 in the product of the 2 polynomials below?
    $(2x^5 + 3x^4 - 9x^3 + 4x + 7)(x^4 - 2x^3 + 5x^2 - 8)$

    F. 0

    G. 1

    H. 2

    J. 8

    K. 9

41. Which of the following is a factor of $6x^2 + x - 35$?

    A. $2x + 5$

    B. $2x - 5$

    C. $3x + 7$

    D. $3x + 5$

    E. $2x + 7$

42. Which of the following is a value of t for $(t - 4)(t + 5) = 0$?

    F. 1

    G. 4

    H. 5

    J. 9

    K. 20

Polynomial Algebra
Advanced Problem Set 4

43. For all $x > 9$, $\dfrac{5x - x^2}{x^2 - 14x + 45}$

    A. $\dfrac{-x}{x - 9}$

    B. $\dfrac{x}{x + 9}$

    C. $\dfrac{1}{x - 9}$

    D. $\dfrac{-1}{45}$

    E. $\dfrac{1}{45}$

44. Which of the following is NOT a solution of $(x + 5)(x - 9)(x - 2)(x + 9) = 0$?

    F. -9

    G. -5

    H. 2

    J. 5

    K. 9

45. What is the value of c if $x + 3$ is a factor of $3x^3 + 3x^2 - 3cx + 9$?

    A. -6

    B. -3

    C. 0

    D. 1

    E. 5

46. $(7a + 2b + c) - (3a + 9b - 2c)$ is equivalent to:

    F. 0

    G. $4a - 7b + 3c$

    H. $4a - 7b - c$

    J. $4a + 11b + 3c$

    K. $4a + 11b - c$

Polynomial Algebra
Advanced Problem Set 4

47. The expression $(5a - 3b)(6a + 7b)$ is equivalent to:

A. $30a^2 - 21b^2$

B. $30a^2 - 53ab - 21b^2$

C. $30a^2 - 53ab + 21b^2$

D. $30a^2 + 17ab - 21b^2$

E. $30a^2 + 17ab + 21b^2$

48. Which of the following expressions is equivalent to $6x^2 - 27x - 105$?

F. $(2x + 5)(x - 7)$

G. $(6x + 5)(x - 7)$

H. $(2x + 5)(3x - 7)$

J. $3(2x + 5)(x - 7)$

K. $3(6x + 5)(x - 7)$

49. For all x, $\dfrac{6x^2 + 3}{3} = ?$

A. $2x^2$

B. $2x^2 + 1$

C. $2x^2 + 3$

D. $6x^2$

E. $6x^2 + 1$

50. What is the sum of the 2 solutions of the equation $x^2 + 2x - 143$?

F. -13

G. -2

H. 0

J. 2

K. 11

Polynomial Algebra
Advanced Problem Set 4

51. For all x, $(4x - 7)^2 = ?$

   A. $8x - 14$

   B. $8x^2 - 14$

   C. $16x^2 + 49$

   D. $16x^2 - 28x + 49$

   E. $16x^2 - 56x + 49$

52. Which of the following quadratic equations has solutions $x = 2a$ and $x = b$?

   F. $x^2 - 2ab = 0$

   G. $x^2 - x(3b - 6a) - 2ab = 0$

   H. $x^2 - x(b + 2a) + 2ab = 0$

   J. $x^2 + x(b - 2a) - 2ab = 0$

   K. $x^2 + x(b + 2a) + 2ab = 0$

DO YOUR FIGURING HERE

Polynomial Algebra
Advanced Problem Set 4

Answer Key

#	Answer	Frequency	Difficulty
1	C	popular	1
2	J	popular	2
3	A	popular	2
4	K	popular	1
5	E	popular	3
6	K	popular	1
7	B	popular	2
8	K	popular	1
9	A	popular	2
10	F	popular	2
11	D	popular	1
12	K	popular	1
13	A	popular	1
14	J	popular	1
15	A	popular	2
16	K	popular	1
17	D	popular	1
18	J	popular	1
19	A	popular	2
20	G	popular	1
21	B	popular	2
22	J	popular	1
23	E	popular	2
24	G	popular	1
25	E	popular	2
26	J	popular	1
27	E	popular	1
28	J	popular	1
29	E	popular	1
30	J	popular	1
31	E	popular	2
32	J	popular	2
33	D	popular	1
34	K	popular	2
35	B	popular	3
36	F	popular	2
37	B	popular	1
38	G	popular	1
39	C	popular	2
40	H	popular	2
41	A	popular	3
42	G	popular	1
43	A	popular	2
44	J	popular	1
45	E	popular	3
46	G	popular	1
47	D	popular	1
48	J	popular	2
49	B	popular	1
50	G	popular	2
51	E	popular	1
52	H	popular	3

Complete the Square
Advanced Problem Set 5

1. $x^2 - 4x + 2$
   Vertex Form:
   Vertex (h, k):
   Sum of the roots:
   Product of the roots:
   Axis of Symmetry:

2. $x^2 + 10x - 4$
   Vertex Form:
   Vertex (h, k):
   Sum of the roots:
   Product of the roots:
   Axis of Symmetry:

3. $x^2 - 9x + 6$
   Vertex Form:
   Vertex (h, k):
   Sum of the roots:
   Product of the roots:
   Axis of Symmetry:

DO YOUR FIGURING HERE

Complete the Square
Advanced Problem Set 5

4. $x^2 - 20x - 4$
   Vertex Form:
   Vertex (h, k):
   Sum of the roots:
   Product of the roots:
   Axis of Symmetry:

5. $x^2 + 12x + 2$
   Vertex Form:
   Vertex (h, k):
   Sum of the roots:
   Product of the roots:
   Axis of Symmetry:

6. $x^2 - 9x + 9$
   Vertex Form:
   Vertex (h, k):
   Sum of the roots:
   Product of the roots:
   Axis of Symmetry:

Complete the Square
Advanced Problem Set 5

7. $x^2 - 7x + 2$
   Vertex Form:
   Vertex (h, k):
   Sum of the roots:
   Product of the roots:
   Axis of Symmetry:

8. $x^2 - 5x + 5$
   Vertex Form:
   Vertex (h, k):
   Sum of the roots:
   Product of the roots:
   Axis of Symmetry:

9. $2x^2 - 4x + 2$
   Vertex Form:
   Vertex (h, k):
   Sum of the roots:
   Product of the roots:
   Axis of Symmetry:

DO YOUR FIGURING HERE

Complete the Square
Advanced Problem Set 5

10. $4x^2 + 8x - 12$
    Vertex Form:
    Vertex (h, k):
    Sum of the roots:
    Product of the roots:
    Axis of Symmetry:

11. $3x^2 + 6x + 2$
    Vertex Form:
    Vertex (h, k):
    Sum of the roots:
    Product of the roots:
    Axis of Symmetry:

12. $4x^2 - 2x - 16$
    Vertex Form=
    Vertex (h, k)=
    Sum of the roots=
    Product of the roots=
    Axis of Symmetry=

Complete the Square
Advanced Problem Set 5

13. $6x^2 + 12x + 2$
    Vertex Form:
    Vertex (h, k):
    Sum of the roots:
    Product of the roots:
    Axis of Symmetry:

14. $9x^2 - 18x + 9$
    Vertex Form:
    Vertex (h, k):
    Sum of the roots:
    Product of the roots:
    Axis of Symmetry:

15. $4x^2 + 12x - 10$
    Vertex Form:
    Vertex (h, k):
    Sum of the roots:
    Product of the roots:
    Axis of Symmetry:

DO YOUR FIGURING HERE

Complete the Square
Advanced Problem Set 5

16. $5x^2 + 15x - 5$
Vertex Form:
Vertex (h, k):
Sum of the roots:
Product of the roots:
Axis of Symmetry:

17. $x^2 + y^2 + 12x - 4y + 24 = 0$
Center (h, k):
Radius:
Standard Form:

18. $9x^2 + 9y^2 + 18x - 180y + 908 = 0$
Center (h, k):
Radius=
Standard Form=

19. $x^2 + y^2 - 4x + 2y + 1 = 0$
Center (h, k):
Radius:
Standard Form:

DO YOUR FIGURING HERE

Complete the Square
Advanced Problem Set 5

20. $x^2 + y^2 - 16x - 6y + 72 = 0$
Center (h, k):
Radius:
Standard Form:

21. $4x^2 + 4y^2 - 40x - 8y + 95 = 0$
Center (h, k):
Radius:
Standard Form:

22. $x^2 + y^2 - 24y = 0$
Center (h, k):
Radius:
Standard Form:

Complete the Square
Advanced Problem Set 5

Answer Key

#	Answer	Frequency	Difficulty
1	Vertex Form: $(x-2)^2 - 2$ Vertex: $(2,-2)$ Sum of the roots: 4 Product of the roots: 4 Axis of Symmetry: $x=2$	popular	1
2	Vertex Form: $(x+5)^2 - 29$ Vertex: $(-5,-29)$ Sum of the roots: -4 Product of the roots: 25 Axis of Symmetry: $x=-5$	popular	1
3	Vertex Form= $\left(x - \frac{9}{2}\right)^2 - \frac{57}{4}$ Vertex= $\left(\frac{9}{2}, -\frac{57}{4}\right)$ Sum of the roots= 9 Product of the roots= 6 Axis of Symmetry $x = \frac{9}{2}$	popular	3
4	Vertex Form: $(x-10)^2 - 104$ Vertex: $(10,-104)$ Sum of the roots: 20 Product of the roots: 100 Axis of Symmetry: $x=10$	popular	2

Complete the Square
Advanced Problem Set 5

#	Answer	Frequency	Difficulty	#	Answer	Frequency	Difficulty
5	Vertex Form: $(x+6)^2 - 34$ Vertex: (-6,-34) Sum of the roots: -12 Product of the roots: 2 Axis of Symmetry: $x = -6$	popular	2	8	Vertex Form: $\left(x - \frac{5}{2}\right)^2 - \frac{5}{4}$ Vertex: $\left(\frac{5}{2}, -\frac{5}{4}\right)$ Sum of the roots: 5 Product of the roots: 5 Axis of Symmetry: $x = \frac{5}{2}$	popular	3
6	Vertex Form: $\left(x - \frac{9}{2}\right)^2 - \frac{45}{4}$ Vertex: $\left(\frac{9}{2}, -\frac{45}{4}\right)$ Sum of the roots: 9 Product of the roots: 9 Axis of Symmetry: $x = \frac{9}{2}$	popular	3	9	Vertex Form: $2(x-1)^2$ Vertex: (1, 0) Sum of the roots: 2 Product of the roots: 1 Axis of Symmetry: $x = 1$	popular	2
7	Vertex Form: $\left(x - \frac{7}{2}\right)^2 - \frac{41}{4}$ Vertex: $\left(\frac{7}{2}, -\frac{41}{4}\right)$ Sum of the roots: 7 Product of the roots: 2 Axis of Symmetry: $x = \frac{7}{2}$	popular	3	10	Vertex Form: $4(x+1)^2 - 16$ Vertex: (-1,-16) Sum of the roots: -2 Product of the roots: -3 Axis of Symmetry: $x = -1$	popular	3

Complete the Square
Advanced Problem Set 5

#	Answer	Frequency	Difficulty
11	Vertex Form: $3(x+1)^2 - 1$ Vertex: (-1,-1) Sum of the roots: -2 Product of the roots: $\frac{-2}{3}$ Axis of Symmetry: $x = -1$	popular	3
12	Vertex Form= $4\left(x - \frac{1}{4}\right)^2 - \frac{65}{4}$ Vertex= $\left(\frac{1}{4}, -\frac{65}{4}\right)$ Sum of the roots= $\frac{1}{2}$ Product of the roots= -4 Axis of Symmetry $x = \frac{1}{4}$	popular	4
13	Vertex Form: $6(x+1)^2 - 4$ Vertex: (-1,-4) Sum of the roots: -2 Product of the roots: $\frac{1}{3}$ Axis of Symmetry: $x = -1$	popular	3

#	Answer	Frequency	Difficulty
14	Vertex Form: $9(x-1)^2$ Vertex: (1, 0) Sum of the roots: 2 Product of the roots: 1 Axis of Symmetry: $x = 1$	popular	2
15	Vertex Form: $4\left(x + \frac{3}{2}\right)^2 - 19$ Vertex: $\left(-\frac{3}{2}, -19\right)$ Sum of the roots: -3 Product of the roots: $-\frac{5}{2}$ Axis of Symmetry: $x = -\frac{3}{2}$	popular	3
16	Vertex Form: $5\left(x + \frac{3}{2}\right)^2 - \frac{65}{4}$ Vertex: $\left(-\frac{3}{2}, -\frac{65}{4}\right)$ Sum of the roots: -3 Product of the roots: -1 Axis of Symmetry: $x = -\frac{3}{2}$	popular	4

Complete the Square
Advanced Problem Set 5

#	Answer	Frequency	Difficulty
17	Center: (-6,2) Radius: 4 Standard Form: $(x+6)^2 + (y-2)^2 = 16$	popular	3
18	Center: (-1,10) Radius= $\frac{1}{3}$ Standard Form= $(x+1)^2 + (y-10)^2 = \frac{1}{9}$	popular	4
19	Center: (8,3) Radius: 2 Standard Form: $(x-2)^2 + (y+1)^2 = 4$	popular	3
20	Center: (8,3) Radius: 1 Standard Form: $(x-8)^2 + (y-3)^2 = 1$	popular	3
21	Center: (5,1) Radius: $\frac{3}{2}$ Standard Form: $(x-5)^2 + (y-1)^2 = \frac{9}{4}$	popular	3
22	Center: (0,12) Radius: 12 Standard Form: $x^2 + (y-12)^2 = 144$	popular	3

Even and Odd Functions
Advanced Problem Set 6

1. The function $f(x)$ is graphed in the xy-plane. Is $f(x)$ even, odd, or neither?

   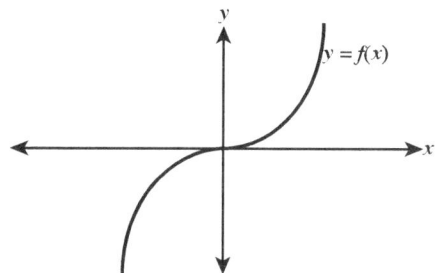

   A. Even
   B. Odd
   C. Neither

2. The function $f(x)$ is graphed in the xy-plane. Is $f(x)$ even, odd, or neither?

   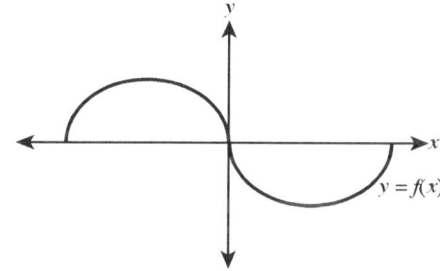

   F. Even
   G. Odd
   H. Neither

3. The function $f(x)$ is graphed in the xy-plane. Is $f(x)$ even, odd, or neither?

   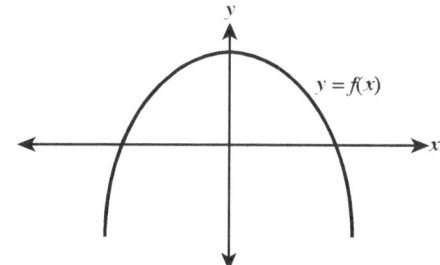

   A. Even
   B. Odd
   C. Neither

Even and Odd Functions
Advanced Problem Set 6

4. The function $f(x)$ is graphed in the xy-plane. Is $f(x)$ even, odd, or neither?

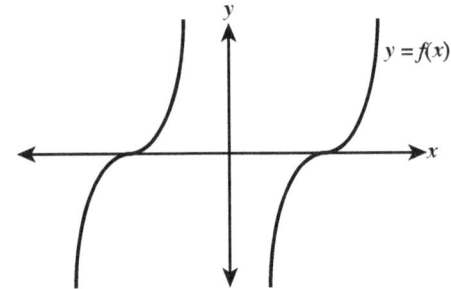

   F. Even

   G. Odd

   H. Neither

5. $f(x) = 3x + 4$
   Is $f(x)$ even, odd, or neither?

   A. Even

   B. Odd

   C. Neither

6. $f(x) = 7x^2$
   Is $f(x)$ even, odd, or neither?

   F. Even

   G. Odd

   H. Neither

7. $f(x) = \cos x$
   Is $f(x)$ even, odd, or neither?

   A. Even

   B. Odd

   C. Neither

8. $f(x) = 2x^3$
   Is $f(x)$ even, odd, or neither?

   F. Even

   G. Odd

   H. Neither

Even and Odd Functions
Advanced Problem Set 6

Answer Key

#	Answer	Frequency	Difficulty
1	B	average	1
2	G	average	1
3	A	average	1
4	H	average	1
5	C	average	1
6	F	average	1
7	A	average	1
8	G	average	1

Conics
Advanced Problem Set 7

1. What are the zeroes of $y = x^2 + 6x - 40$?
   A. $(-4, 10)$
   B. $(4, -10)$
   C. $(-5, 8)$
   D. $(5, -8)$
   E. $(-4, 2)$

2. Which of the following is a positive root of $x^2 + \frac{53}{30}x - \frac{7}{10}$?
   F. $\frac{1}{3}$
   G. $\frac{3}{5}$
   H. $1\frac{1}{10}$
   J. $1\frac{2}{5}$
   K. $2\frac{1}{10}$

3. If a circle with its center at the origin contains the point $\left(-5, 2\sqrt{6}\right)$, what is an equation for this circle?
   A. $x^2 + y^2 = 7$
   B. $x^2 + y^2 = 25$
   C. $\frac{x^2}{25} + \frac{y^2}{24} = 1$
   D. $x^2 + y^2 = 49$
   E. $x^2 + y^2 = 625$

4. What are the coordinates of the foci of the hyperbola $4x^2 - 25y^2 = 100$?
   F. $(4.58, 0), (-4.58, 0)$
   G. $(5.39, 0), (-5.39, 0)$
   H. $(0, 21), (0, -21)$
   J. $(0, 4.85), (0, -4.85)$
   K. $(0, 5.39), (0, -5.39)$

Conics
Advanced Problem Set 7

5. What is the center of an ellipse described by $16x^2 + 4y^2 - 64x + 32y = 64$?

    A. $(8, -4)$

    B. $(2, -4)$

    C. $(4, 4)$

    D. $(-2, 4)$

    E. $(-4, 2)$

6. If $f(x) = ax^2 + bx + c$, what is $a + b$ if $f(0) = 4$ and $f(1) = 12$?

    F. 3

    G. 4

    H. 6

    J. 8

    K. 12

7. Where is the center of an ellipse with the equation $\dfrac{(x+3)^2}{36} + \dfrac{(y-4)^2}{9} = 1$?

    A. $(6, 3)$

    B. $(3, 6)$

    C. $(3, 4)$

    D. $(-3, 4)$

    E. $(3, -4)$

8. What is the equation of a circle with a radius 6.5 centered at $(-4, 5)$?

    F. $(x-4)^2 + (y+5)^2 = 42.25$

    G. $(x+4)^2 + (y-5)^2 = 42.25$

    H. $(x+2)^2 - (y-5)^2 = 42.25$

    J. $(x-5)^2 + (y+4)^2 = 6.5$

    K. $(x+5)^2 + (y-4)^2 = 6.5$

DO YOUR FIGURING HERE

Conics
Advanced Problem Set 7

9. What are the x-intercepts of the hyperbola with the equation $\dfrac{(x+3)^2}{5} - \dfrac{y^2}{16} = 1$?

   A. $(-3, 0), (3, 0)$
   B. $(-5, 0), (5, 0)$
   C. $(-4, 0), (4, 0)$
   D. $\left(3 - \sqrt{5}, 0\right), \left(3 + \sqrt{5}, 0\right)$
   E. $\left(-3 - \sqrt{5}, 0\right), \left(-3 + \sqrt{5}, 0\right)$

10. If the roots of a polynomial function are -2 and 7, what is a possible equation for the function?

    F. $y = x^2 - 2x + 7$
    G. $x^2 - 14x + 5$
    H. $x^2 + 14x + 5$
    J. $x^2 - 5x - 14$
    K. $x^2 + 5x - 14$

11. The number of points that satisfy both $x^2 + y^2 = 36$ and $y = x^2 - 8x + 16$ is:

    A. 0
    B. 1
    C. 2
    D. 3
    E. 4

12. What are the coordinates of the foci of the hyperbola $4y^2 - 8x^2 = 56$?

    F. $(2, 0), (-2, 0)$
    G. $(4.58, 0), (-4.58, 0)$
    H. $(0, 2), (0, -2)$
    J. $(0, 4.58), (0, -4.58)$
    K. $(0, 4), (0, -4)$

Conics
Advanced Problem Set 7

13. If the roots of a quadratic function are $(3 + 2i)$ and $(3 - 2i)$ what is the possible equation for this function?

    A. $y = x^2 + 6x + 13$

    B. $y = x^2 - 6x + 13$

    C. $y = x^2 + 6x - 13$

    D. $y = x^2 - 6x - 13$

    E. $y = -x^2 - 6x + 13$

14. If $x^2 + 16x - 36 = K(x - 2)$, what is K?

    F. $x - 20$

    G. $x - 18$

    H. $x - 4$

    J. $x + 6$

    K. $x + 18$

15. Where is the center of a circle with the equation $(x - 4) + (y + 3)^2 = 62$?

    A. $(4, 3)$

    B. $(-4, -3)$

    C. $(4, -3)$

    D. $(3, 4)$

    E. $(-3, 4)$

16. What is the maximum value of the function $y = -x^2 - 10x - 24$?

    F. -34

    G. -24

    H. -14

    J. -5

    K. 1

Conics
Advanced Problem Set 7

17. What is the minimum value of the function $x^2 - 11x + 24$?

    A. -11
    B. -6.25
    C. 3
    D. 5.5
    E. 8

18. If $4x^2 + kx + 25$ yields two real, unequal roots, what are all possible values of k?

    F. $k = 10$
    G. $k > 10$
    H. $k < 5$
    J. $k > 20$ or $k < -20$
    K. $k \geq 10$

19. What value of k would yield two real, equal roots for the function $kx^2 - 6x + 3$?

    A. -6
    B. -3
    C. 1.5
    D. 2
    E. 3

20. $f(x) = 6x^2 + 9x - 3$ has roots that are?

    F. real, equal, and rational
    G. real, equal, and irrational
    H. real, unequal, and rational
    J. real, unequal, and irrational
    K. imaginary

Conics
Advanced Problem Set 7

21. Where is the center of a circle with the equation $2x^2 - 8x + 2y^2 + 16y + 40 = 36$?

    A. $(2, -4)$

    B. $(-2, 4)$

    C. $(-2, -4)$

    D. $(-4, 10)$

    E. $(-5, 8)$

22. What are the coordinates of the center of the ellipse given by the equation $3x^2 - 18x + 4y^2 + 16y = -20$?

    F. $(-3, 2)$

    G. $(3, -2)$

    H. $(2, -3)$

    J. $(-2, 3)$

    K. $(3, 2)$

23. The equation $y^2 - x^2 - 6y + 2x - 10 = 0$ is centered at which of the following points?

    A. $(-1, -3)$

    B. $(1, 3)$

    C. $(1, 0)$

    D. $(3, 1)$

    E. $(-3, -1)$

24. What is the sum of the roots of $y = \sqrt{7}x^2 + \sqrt{11}x + 1.85$?

    F. -1.571

    G. -1.254

    H. $-.627$

    J. $.699$

    K. 1.107

Conics
Advanced Problem Set 7

25. What is the sum of the roots of $y = 2\sqrt{5}x^2 - 6\sqrt{7}x + 0.8$?

    A. -15.08

    B. -11.4

    C. -.179

    D. 3.45

    E. 3.55

26. The graph of $9x^2 - 18x + 9y^2 + 36y + 36 = 1$ describes?

    F. 2 lines

    G. A parabola

    H. A circle

    J. An ellipse

    K. A hyperbola

27. What is the length of the major axis of this ellipse $3(x+4)^2 + (y+2)^2 = 7$?

    A. 1.52

    B. 2.65

    C. 3.05

    D. 5.29

    E. 7.00

28. What is the product of the roots of $y = 3.7x^2 + \sqrt{5}x - 12$?

    F. -6.27

    G. -3.24

    H. -.604

    J. 1.46

    K. 3.65

DO YOUR FIGURING HERE

Conics
Advanced Problem Set 7

29. What is the product of the roots of $y = -0.7x^2 + \sqrt{6}x + 4\sqrt{2}$?

    A. -8.08
    B. -3.95
    C. -.29
    D. 3.5
    E. 22.36

30. What is the area of the figure enclosed by the equation $x^2 - 6x + y^2 - 10y = 30$?

    F. 16π
    G. 30π
    H. 36π
    J. 64π
    K. 100π

31. What are the equations of the asymptotes of $16x^2 - 25y^2 = 400$?

    A. $\pm \dfrac{25}{16} x = y$
    B. $\pm \dfrac{4}{5} x = y$
    C. $\pm \dfrac{5}{4} x = y$
    D. $\pm \dfrac{16}{25} x = y$
    E. $\pm \dfrac{1}{16} x = y$

32. The sum of the two roots of a quadratic equation is -2 and their product is -30. Which of the following could be the equation?

    F. $x^2 + 2x - 30 = y$
    G. $x^2 - 2x + 30 = y$
    H. $x^2 - 2x - 30 = y$
    J. $x^2 + 2x + 30 = y$
    K. $x^2 + 30x - 2 = y$

Conics
Advanced Problem Set 7

33. What is a possible equation for a parabola for which the sum of the roots is -3 and the product of the roots is $\frac{7}{3}$?

    A. $y = 3x^2 - 6x + 14$
    B. $y = x^2 - 3x + 7$
    C. $y = x^2 - 4x + 21$
    D. $y = x^2 - 6x + 21$
    E. $y = 3x^2 + 9x + 7$

34. $3x^2 - 12x + 3y^2 + 24y = 6$ gives the graph a(n):

    F. Circle
    G. Parabola
    H. Hyperbola
    J. Ellipse
    K. Line

35. A circle with the equation $x^2 + (y-3)^2 = 15$ is externally tangent to a circle with the equation $x^2 + (y+5)^2 = K$. What is the value of K?

    A. 3.873
    B. 4.127
    C. 8.254
    D. 15.99
    E. 17.032

36. The equation $xy = -25$ contains points in which of the following quadrants?

    F. I only
    G. I and III only
    H. II and III only
    J. II and IV only
    K. I, II, III, and IV

Conics
Advanced Problem Set 7

Answer Key

#	Answer	Frequency	Difficulty
1	B	rare	1
2	F	rare	3
3	D	rare	2
4	G	rare	3
5	B	rare	3
6	J	rare	2
7	D	rare	1
8	G	rare	1
9	E	rare	2
10	J	rare	2
11	C	rare	3
12	J	rare	4
13	B	rare	3
14	K	rare	3
15	C	rare	1
16	K	rare	3
17	B	rare	3
18	J	rare	2
19	E	rare	2
20	J	rare	2
21	A	rare	3
22	G	rare	3
23	B	rare	3
24	G	rare	4
25	E	rare	4
26	H	rare	3
27	D	rare	3
28	G	rare	2
29	A	rare	2
30	J	rare	2
31	B	rare	2
32	F	rare	3
33	E	rare	4
34	F	rare	3
35	E	rare	4
36	J	rare	2

Ellipses
Advanced Problem Set 8

1. Sketch the graph of each ellipse. Identify the center, foci, vertices, and the co-vertices.
$$\frac{x^2}{25} + \frac{y^2}{4} = 1$$

2. Sketch the graph of each ellipse. Identify the center, foci, vertices, and the co-vertices.
$$\frac{x^2}{1} + \frac{y^2}{9} = 1$$

3. Sketch the graph of each ellipse. Identify the center, foci, vertices, and the co-vertices.
$$\frac{x^2}{4} + \frac{y^2}{9} = 1$$

4. Sketch the graph of each ellipse. Identify the center, foci, vertices, and the co-vertices.
$$\frac{(x+2)^2}{4} + \frac{(y+1)^2}{9} = 1$$

DO YOUR FIGURING HERE

Ellipses
Advanced Problem Set 8

5. Sketch the graph of each ellipse. Identify the center, foci, vertices, and the co-vertices.

$$\frac{(x-2)^2}{9} + \frac{(y-2)^2}{4} = 1$$

6. Sketch the graph of each ellipse. Identify the center, foci, vertices, and the co-vertices.

$$\frac{x^2}{1} + \frac{(y+2)^2}{9} = 1$$

7. Sketch the graph of each ellipse. Identify the center, foci, vertices, and the co-vertices.

$$\frac{(x-1)^2}{4} + \frac{(y-3)^2}{4} = 1$$

8. Sketch the graph of each ellipse. Identify the center, foci, vertices, and the co-vertices.

$$16(x+1)^2 + 9(y-1)^2 = 144$$

DO YOUR FIGURING HERE

Ellipses
Advanced Problem Set 8

9. Sketch the graph of each ellipse. Identify the center, foci, vertices, and the co-vertices.
$9(x-1)^2 + 25(y+2)^2 = 225$

10. Sketch the graph of each ellipse. Identify the center, foci, vertices, and the co-vertices.
$4x^2 + 25y^2 = 100$

11. Write the standard equation for the ellipse with the given characteristics.
Foci:$(5,0)$, $(-5,0)$
Vertices:$(9,0)$, $(-9,0)$

12. Write the standard equation for the ellipse with the given characteristics.
Foci:$(7,0)$, $(-7,0)$
Co-Vertices:$(0,3)$, $(0,-3)$

Ellipses
Advanced Problem Set 8

13. Write the standard equation for the ellipse with the given characteristics.
    Vertices: $(5, 0)$, $(-5, 0)$
    Co-Vertices: $(0, 4)$, $(0, -4)$

14. Write the standard equation for the ellipse with the given characteristics.
    The major axis is 16 units long and parallel to the x-axis. The center is $(5, 4)$ and the minor axis is 9 units long.

15. Write the standard equation for the ellipse with the given characteristics.
    The endpoints of the major axis are at $(2, 12)$ and $(2, -4)$. The endpoints of the minor axis are at $(4, 4)$ and $(0, 4)$.

16. Write the standard equation for the ellipse with the given characteristics.
    The major axis is 12 units long and parallel to the y-axis. The minor axis is 8 units long and the center is at $(-2, 3)$.

DO YOUR FIGURING HERE

Ellipses
Advanced Problem Set 8

17. Write the standard equation for the ellipse with the given characteristics.
    The endpoints of the minor axis are at (-2, 5) and (-2, -1). The endpoints of the major axis are at (-9, 2) and (5, 2).

18. Write the standard equation for each ellipse. Identify the coordinates of the center, vertices, co-vertices, and foci.
    $x^2 + 4y^2 + 6x - 8y = 3$

19. Write the standard equation for each ellipse. Identify the coordinates of the center, vertices, co-vertices, and foci.
    $16x^2 + 4y^2 + 32x - 8y = 44$

20. Write the standard equation for each ellipse. Identify the coordinates of the center, vertices, co-vertices, and foci.
    $x^2 + 16y^2 - 64y = 0$

DO YOUR FIGURING HERE

Ellipses
Advanced Problem Set 8

21. Write the standard equation for each ellipse. Identify the coordinates of the center, vertices, co-vertices, and foci.
    $25x^2 + y^2 - 50x = 0$

22. Write the standard equation for each ellipse. Identify the coordinates of the center, vertices, co-vertices, and foci.
    $4x^2 + 9y^2 - 16x + 18y = 11$

23. Write the standard equation for each ellipse. Identify the coordinates of the center, vertices, co-vertices, and foci.
    $25x^2 + 9y^2 + 100x + 18y = 116$

24. Write the standard equation for each ellipse. Identify the coordinates of the center, vertices, co-vertices, and foci.
    $9x^2 + 16y^2 - 36x - 64y - 44 = 0$

Ellipses
Advanced Problem Set 8

25. Write the standard equation for each ellipse. Identify the coordinates of the center, vertices, co-vertices, and foci.
$36x^2 + 25y^2 - 72x + 100y = 764$

26. Write the standard equation for each ellipse. Identify the coordinates of the center, vertices, co-vertices, and foci.
$7x^2 + 3y^2 - 28x - 12y + 19 = 0$

27. Write the standard equation for each ellipse. Identify the coordinates of the center, vertices, co-vertices, and foci.
$16x^2 + 25y^2 + 32x - 150y = 159$

28. Mars orbits the Sun in an elliptical path whose minimum distance from the Sun is 129.5 miles and whose maximum distance from the Sun is 154.4 million miles. The Sun represents one focus of the ellipse. Write the standard equation for the elliptical orbit of Mars around the Sun, where the center of the ellipse is at the origin.

Ellipses
Advanced Problem Set 8

29. A satellite is in an elliptical orbit with the center of Earth at one focus. The major axis of the orbit is 28,900 miles long and the center of the Earth is 8000 miles from the center of the ellipse. Assuming that the center of the ellipse is the origin and the foci lie on the x-axis, write the equation of the path of the satellite.

30. The moon orbits Earth in an elliptical path with the center of the Earth at one focus. The major axis of the orbit is 774,000 kilometers, and the minor axis is 773,000 kilometers. Using $(0,0)$ as the center of the ellipse, write the standard equation for the orbit of the Moon around the Earth. How far from the center of Earth is the Moon at its closest point? How far from the center of Earth is the Moon at its farthest point?

Ellipses
Advanced Problem Set 8

Answer Key

#	Answer	Frequency	Difficulty
1	Center=$(0,0)$ Foci=$\left(-\sqrt{21},0\right)$, $\left(\sqrt{21},0\right)$ Vertices=$(-5,0)$, $(5,0)$ Co-Vertices=$(-2,0)$, $(2,0)$	popular	1
2	Center=$(0,0)$ Foci=$\left(0,-2\sqrt{2}\right)$, $\left(0,2\sqrt{2}\right)$ Vertices=$(0,-3)$, $(0,3)$ Co-Vertices=$(-1,0)$, $(1,0)$	popular	1
3	Center=$(0,0)$ Foci=$\left(0,-\sqrt{5}\right)$, $\left(0,\sqrt{5}\right)$ Vertices=$(0,-3)$, $(0,3)$ Co-Vertices=$(-2,0)$, $(2,0)$	popular	1
4	Center=$(-2,-1)$ Foci=$\left(-2,-1-\sqrt{5}\right)$, $\left(-2,-1+\sqrt{5}\right)$ Vertices=$(-2,-4)$, $(-2,2)$ Co-Vertices=$(0,-1)$, $(-4,-1)$	popular	1
5	Center=$(2,2)$ Foci=$\left(2-\sqrt{5},2\right)$, $\left(2+\sqrt{5},2\right)$ Vertices=$(-1,2)$, $(5,2)$ Co-Vertices=$(2,0)$, $(2,4)$	popular	1
6	Center=$(0,-2)$ Foci=$\left(0,-2-2\sqrt{2}\right)$, $\left(0,-2+2\sqrt{2}\right)$ Vertices=$(0,-5)$, $(0,1)$ Co-Vertices=$(1,-2)$, $(-1,-2)$	popular	1

Ellipses
Advanced Problem Set 8

#	Answer	Frequency	Difficulty
7	Center=(1, 3) Foci=(1, 3) Vertices=(1, 1), (1, 5), (-1, 3), (3, 3)	popular	1
8	Center=(-1, 1) Foci=$\left(-1, 1 - \sqrt{7}\right)$, $\left(-1, 1 + \sqrt{7}\right)$ Vertices=(-1, -3), (-1, 5) Co-Vertices=(-4, 1), (2, 1)	popular	2
9	Center=(1, -2) Foci=(-3, -2), (5, -2) Vertices=(-4, -2), (6, -2) Co-Vertices=(1, -5), (1, 1)	popular	2
10	Center=(0, 0) Foci=$\left(-\sqrt{21}, 0\right)$, $\left(\sqrt{21}, 0\right)$ Vertices=(-5, 0), (5, 0) Co-Vertices=(0, -2), (0, 2)	popular	2
11	Equation: $\dfrac{x^2}{81} + \dfrac{y^2}{56} = 1$	popular	3
12	Equation: $\dfrac{x^2}{58} + \dfrac{y^2}{9} = 1$	popular	3
13	Equation: $\dfrac{x^2}{25} + \dfrac{y^2}{16} = 1$	popular	3
14	Equation: $\dfrac{(x-5)^2}{64} + \dfrac{(y-4)^2}{20.25} = 1$	popular	3
15	Equation: $\dfrac{(x-2)^2}{4} + \dfrac{(y-4)^2}{64} = 1$	popular	3
16	Equation: $\dfrac{(x+2)^2}{16} + \dfrac{(y-3)^2}{36} = 1$	popular	3
17	Equation: $\dfrac{(x+2)^2}{49} + \dfrac{(y-2)^2}{9} = 1$	popular	3

Ellipses
Advanced Problem Set 8

#	Answer	Frequency	Difficulty
18	Equation: $\frac{(x+3)^2}{16} + \frac{(y-1)^2}{8} = 1$ Center: $(-3, 1)$ Vertices: $(-7, 1), (1, 1)$ Co-Vertices: $\left(-3, 1 - 2\sqrt{2}\right), \left(-3, 1 + 2\sqrt{2}\right)$ Foci: $\left(-3 - 2\sqrt{2}, 1\right), \left(-3 + 2\sqrt{2}, 1\right)$	popular	4
19	Equation: $\frac{(x+1)^2}{4} + \frac{(y-1)^2}{16} = 1$ Center: $(-1, 1)$ Vertices: $(-1, -3), (-1, 5)$ Co-Vertices: $(1, 1), (-3, 1)$ Foci: $\left(-1, 1 - 2\sqrt{3}\right), \left(-1, 1 + 2\sqrt{3}\right)$	popular	4
20	Equation: $\frac{x^2}{64} + \frac{(y-2)^2}{4} = 1$ Center: $(0, 2)$ Vertices: $(-8, 2), (8, 2)$ Co-Vertices: $(0, 0), (0, 4)$ Foci: $\left(-2\sqrt{15}, 2\right), \left(2\sqrt{15}, 2\right)$	popular	4

Ellipses
Advanced Problem Set 8

#	Answer	Frequency	Difficulty
21	Equation: $(x-1)^2 + \dfrac{y^2}{25} = 1$ Center: $(1, 0)$ Vertices: $(1, -5), (1, 5)$ Co-Vertices: $(0, 0), (2, 0)$ Foci: $\left(1, -2\sqrt{6}\right), \left(1, 2\sqrt{6}\right)$	popular	4
22	Equation: $\dfrac{(x-2)^2}{9} + \dfrac{(y+1)^2}{4} = 1$ Center: $(2, -1)$ Vertices: $(-1, -1), (5, -1)$ Co-Vertices: $(2, -3), (2, 1)$ Foci: $\left(2 - 2\sqrt{5}, -1\right), \left(2 + \sqrt{5}, -1\right)$	popular	4
23	Equation: $\dfrac{(x+2)^2}{9} + \dfrac{(y+1)^2}{25} = 1$ Center: $(-2, -1)$ Vertices: $(-2, -6), (-2, 4)$ Co-Vertices: $(-5, -1), (1, -1)$ Foci: $(-2, -5), (-2, 3)$	popular	4

Ellipses
Advanced Problem Set 8

#	Answer	Frequency	Difficulty
24	Equation: $\frac{(x-2)^2}{16} + \frac{(y-2)^2}{9} = 1$ Center: (2, 2) Vertices: (-2, 2), (6, 2) Co-Vertices: (2, -1), (2, 5) Foci: $(2 - \sqrt{7}, 2)$, $(2 + \sqrt{7}, 2)$	popular	4
25	Equation: $\frac{(x-1)^2}{25} + \frac{(y+2)^2}{36} = 1$ Center: (1, -2) Vertices: (1, 4), (1, -8) Co-Vertices: (6, -2), (-4, -2) Foci: $(1, -2 - \sqrt{11})$, $(1, -2 + \sqrt{11})$	popular	4
26	Equation: $\frac{(x-2)^2}{3} + \frac{(y-2)^2}{7} = 1$ Center: (2, 2) Vertices: $(2, 2 - \sqrt{7})$, $(2, 2 + \sqrt{7})$ Co-Vertices: $(2 - \sqrt{3}, 2)$, $(2 + \sqrt{3}, 2)$ Foci: (2, 0), (2, 4)	popular	4

Ellipses
Advanced Problem Set 8

#	Answer	Frequency	Difficulty
27	Equation: $\frac{(x+2)^2}{28} + \frac{(y-3)^2}{17.92} = 1$ Center: $(-2, 3)$ Vertices: $\left(-2 - 2\sqrt{7}, 3\right), \left(-2 + 2\sqrt{7}, 3\right)$ Co-Vertices: $\left(-2, 3 - \sqrt{17.92}\right), \left(-2, 3 + \sqrt{17.92}\right)$ Foci: $\left(-2 - \sqrt{10.08}, 3\right), \left(-2 + \sqrt{10.08}, 3\right)$	popular	4
28	Equation: $\frac{x^2}{20149.8} + \frac{y^2}{19994.8} = 1$	popular	3
29	Equation: $\frac{x^2}{2.09 * 10^8} + \frac{y^2}{1.45 * 10^8} = 1$	popular	3
30	Furthest Point = 406,666 km Closest Point = 367,334 km	popular	3

Logarithms
Quick Drill

1. Rewrite in exponential form:
$\log_u x = v$

2. Rewrite in exponential form:
$\log_8 x = y$

3. Rewrite in exponential form:
$\log_8 b^3 = a$

4. Rewrite in exponential form:
$\log_4 (xy) = 3$

5. Rewrite in logarithmic form:
$6^{xy} = z$

6. Rewrite in logarithmic form:
$j^{-1} = k$

7. Rewrite in logarithmic form:
$16^r = q$

8. Rewrite in logarithmic form:
$x^{y^2} = 3$

9. Solve for x:
$\log_{10} 1000 = x$

10. Solve for x:
$\log_3 9 = x$

11. Solve for x:
$\log_x 8 = 3$

12. Solve for x:
$\log_4 x = 4$

13. Solve for x:
$\log_4 (x^2) = 4$

Logarithms
Quick Drill

Answer Key

#	Answer
1	$u^v = x$
2	$8^y = x$
3	$8^a = b^3$
4	$4^3 = xy$ or $64 = xy$
5	$\log_6 z = xy$
6	$\log_j k = -1$
7	$\log_{16} q = r$
8	$\log_x 3 = y^2$
9	$x = 3$
10	$x = 2$
11	$x = 2$
12	$x = 256$
13	$x = 16$

Logs
Advanced Problem Set 9

1. If $\log_x 49 = 2$, then $x =$

2. If $\log_2 16 = x$, then $x =$

3. If $\log_3 x = 8$, then $x =$

4. If $\log x^2 = 2$, then $x =$

5. If $\log 100 + \log 100 = x$, then $x =$

6. If $\log_3 x^2 - \log_3 x = 2$, then $x =$

7. If $\log_4 x^2 + \log_4 y^2 = 3$, then $xy =$

8. If $\log_a 36 = 2$, then a is equal to:
   F. 2
   G. 4
   H. 6
   J. 18
   K. 34

DO YOUR FIGURING HERE

Logs
Advanced Problem Set 9

9. If $\log_3 x = 3$, then x^2 is equal to:
   A. 6
   B. 9
   C. 27
   D. 81
   E. 729

10. If $a^x = 81$, then $\log_a 81 = x$. What is $\log_a 81^2$ equal to?
    F. 2
    G. x
    H. $2x$
    J. x^2
    K. $(2x)^2$

11. If $\log_x 2 + \log_x 8 = 2$, then x is equal to:
    A. 2
    B. 4
    C. 6
    D. 16
    E. 32

12. If $\log_2 (2x) + \log_2 (2x) = 5$, then x is equal to:
    F. $\sqrt{8}$
    G. 4
    H. $\sqrt{32}$
    J. 8
    K. 16

Logs
Advanced Problem Set 9

13. If $\log_3 x - \log_3(2x^2) = 3$, then x is equal to:

  A. $\dfrac{1}{54}$

  B. $\dfrac{1}{27}$

  C. 0

  D. 27

  E. 54

DO YOUR FIGURING HERE

14. $\log_x x^2 - \log_x x =$

  F. -1

  G. 0

  H. 1

  J. 2

  K. 3

15. If $\log_x a - \log_x b = 0$, then $\dfrac{a}{b}$ is equal to:

  A. 0

  B. 1

  C. $a - b$

  D. $a + b$

  E. $a^2 + b^2$

16. If $2^x = y$, then $\log_2 y^2$ is equal to:

  F. 0

  G. 1

  H. $2x$

  J. $2y$

  K. y^2

Logs
Advanced Problem Set 9

17. $\log_3 81\sqrt{3} =$
    A. 2
    B. 3
    C. $\dfrac{7}{2}$
    D. $\dfrac{9}{2}$
    E. 81

18. $\log_2 64\sqrt[3]{2} =$
    F. 2
    G. 4
    H. $\dfrac{16}{3}$
    J. 6
    K. $\dfrac{19}{3}$

19. If $\log_2 x + 3\log_2 x = 5$, then $x =$
    A. 2
    B. 2.4
    C. 10.7
    D. 16
    E. 32

20. For all x such that $x > 0$, $f(x) = \log_4 x$. What does $f^{-1}(x)$ equal?
    F. $x^{\frac{1}{4}}$
    G. $4^{\frac{1}{x}}$
    H. 4^x
    J. x^4
    K. $\log_x 4$

Logs
Advanced Problem Set 9

21. If $3\log_3 x - \log_3 x = 10$, then $x =$
   A. 1.78
   B. 3.16
   C. 6.24
   D. 15.59
   E. 243

22. $f(x) = 8^x$ for all real values of x. If $a > 1$ and $b > 1$, then $\dfrac{f^{-1}(a)}{f^{-1}(b)} =$
   F. $\log_a 8 \log_b 8$
   G. $\log_8 a - \log_8 b$
   H. $\dfrac{\log_8 a}{\log_8 b}$
   J. $\log_8 \dfrac{a}{b}$
   K. 8^{a-b}

23. If $\log_a 2 = 16$, then $a =$
   A. 1.04
   B. 1.09
   C. 2.5
   D. 4
   E. 8

24. If $f(x) = e^{\frac{x}{2}}$ and $g(x) = x^{-2}$, then $f(g(2)) =$
   F. 0.07
   G. 0.14
   H. 1.13
   J. 1.65
   K. 2.72

Logs
Advanced Problem Set 9

25. If $f(x) = e^x$ and $g(x) = \dfrac{x^2}{4}$, then $g(f(2)) =$

A. 1.4

B. 3.7

C. 13.6

D. 54.5

E. 745.2

26. If $f(x) = 4\ln x + 5$ and $g(x) = e^x$, then $f(g(3)) =$

F. 1.1

G. 9.4

H. 10.2

J. 17.0

K. 20.1

27. When a certain radioactive element decays, the amount that exists at any time t can be calculated by the function $E(t) = ae^{\frac{t}{600}}$, where a is the initial amount and t is the elapsed time in years. How many years would it take for 900 milligrams of the element to decay to 450 milligrams?

A. 0.5

B. 200.82

C. 415.89

D. 545.88

E. 3665.55

28. If $9.4^x = 4.1^y$, then $\dfrac{y}{x} =$

F. 0.360

G. 0.436

H. 0.630

J. 1.588

K. 2.293

Logs
Advanced Problem Set 9

29. If $f(x) = \log_5 x$ for $x > 0$, then $f^{-1}(x) =$

   A. $\dfrac{5}{x}$

   B. $\dfrac{x}{5}$

   C. $\log_x 5$

   D. x^5

   E. 5^x

30. If $6^{a+3} = 8^a$, then $a =$

   F. $-.037$

   G. 0.86

   H. 2.6

   J. 3.5

   K. 18.7

31. If $f(x) = \log_5 \dfrac{x}{9}$ for $x \geq 9$, then $f^{-1}(x) =$

   A. $5^{\frac{x}{9}}$

   B. $5(9^x)$

   C. $9(5^x)$

   D. $\log_5 \dfrac{x}{9}$

   E. $\dfrac{x}{9}$

32. If $(8.04)^a = (4.33)^b$, what is the value of $\dfrac{a}{b}$?

   F. -0.27

   G. 0.27

   H. 0.57

   J. 0.70

   K. 1.42

Logs
Advanced Problem Set 9

33. If $\log_6 (x^2 + 4) = 2$, then which of the following could be the value of x?

    A. 5.7
    B. 6.3
    C. 7.7
    D. 32
    E. 40

34. If $\log_3 (x^2 - 5) = 4$, which of the following could be the value of x?

    F. 4.12
    G. 7.68
    H. 8.31
    J. 9.27
    K. 11

35. If $f(x) = \log 10^{\frac{x}{2}}$, what is the smallest possible value of x such that $f(x) > 50$?

    A. 26
    B. 49
    C. 51
    D. 99
    E. 101

36. If $\log_3 (x - 15) = \log_9 (x - 3)$, which of the following could be the value of x?

    F. -12
    G. -3
    H. 12
    J. 18
    K. 19

Logs
Advanced Problem Set 9

37. If $\log_4(x-4) = \log_{16}(2x-9)$, then $x =$

   A. -5

   B. -1

   C. 2.6

   D. 5

   E. 13

38. If $\log_x 4 = a$ and $\log_x 3 = b$, then $\log_x 48 =$

   F. $2a+b$

   G. a^2+b

   H. $a^2 b$

   J. $2ab$

   K. $4a+b$

39. If $f(x) = 3^x$, then $f\left(\log_4 \dfrac{1}{256}\right) =$

   A. -12

   B. -4

   C. $\dfrac{1}{256}$

   D. $\dfrac{1}{81}$

   E. 81

Logs
Advanced Problem Set 9

Answer Key

#	Answer	Frequency	Difficulty
1	7	average	2
2	4	average	2
3	6561	average	2
4	10	average	2
5	4	average	2
6	9	average	2
7	8	average	3
8	H	average	3
9	E	average	3
10	H	average	4
11	B	average	3
12	F	average	3
13	A	average	3
14	H	average	3
15	B	average	3
16	H	average	4
17	D	average	3
18	K	average	3
19	B	average	2
20	H	average	2
21	E	average	2
22	H	average	2
23	A	average	2
24	H	average	3
25	C	average	3
26	J	average	3
27	C	average	2
28	J	average	3
29	E	average	2
30	K	average	4
31	C	average	3
32	J	average	3
33	A	average	2
34	J	average	2
35	E	average	3
36	K	average	4
37	D	average	4
38	F	average	3
39	D	average	3

Sequences
Advanced Problem Set 10

1. If the first four terms of a sequence are -8, -5, -2, and 1. What is the 91st term?

    A. 259
    B. 262
    C. 265
    D. 268
    E. 271

2. What is the sum of the first 92 terms of the sequence from question 1?

    F. 11,822
    G. 11,830
    H. 15,480
    J. 16,016
    K. 16,554

3. What is the result when you subtract the sum of the integers from 1 to 30, inclusive, from the sum of the integers from 91 to 130, inclusive?

    A. 452
    B. 955
    C. 985
    D. 3955
    E. 3975

4. If each term in a sequence is 4 more than the previous term and the 65th term is 257, what is the first term?

    F. -4
    G. -3
    H. 1
    J. 2
    K. 4

Sequences
Advanced Problem Set 10

5. If the series $[2 + 4 + 6 + \ldots + x] = 600$, what is the value of x?

   A. 24
   B. 46
   C. 48
   D. 50
   E. 52

6. The fourth term of an arithmetic sequence is 9 more than the first term. If the sum of these two terms is 1, then what is the 10th term of the sequence?

   F. 17
   G. 20
   H. 22
   J. 23
   K. 26

7. The first term of an arithmetic sequence is 4, and the second term is $\frac{1}{3}$ of the sixth term. What is the fourth term in the sequence?

   A. 1.3
   B. 7
   C. 12
   D. 16
   E. 20

8. The 50th term of an arithmetic sequence is 168.5 and the 100th term is 333.5. What is the 199th term?

   F. 600.5
   G. 659.7
   H. 660.2
   J. 699.5
   K. 700.0

Sequences
Advanced Problem Set 10

9. What is the sum of the infinite geometric series $\frac{1}{2} + \frac{1}{4} + \frac{1}{8} + \frac{1}{16} + \ldots$?

   A. $\frac{1}{2}$

   B. 1

   C. $\frac{3}{2}$

   D. 2

   E. $\frac{5}{2}$

10. The front row in a theater has 10 seats. Each of the remaining rows has 2 more seats than the row in front of it. If the theater has 26 rows, how many seats does it have?

    F. 360

    G. 850

    H. 910

    J. 936

    K. 1560

11. $\sum\limits_{n=1}^{75}(n-5) =$

    A. 1762.5

    B. 2437.0

    C. 2475.0

    D. 2625.0

    E. 2812.5

12. If $\sum\limits_{n=4}^{80}(n+5) = x + \sum\limits_{n=1}^{77}(n+5)$, what is the value of x?

    F. 192

    G. 228

    H. 231

    J. 462

    K. 638

DO YOUR FIGURING HERE

Sequences
Advanced Problem Set 10

13. The first three terms of an arithmetic sequence are $4q - 3$, $6q$, and $6q - 3$ for some real number q. What is the value of the fourth term of this sequence?

    A. -27
    B. -24
    C. -18
    D. 18
    E. 24

14. For some real number q, the first three terms of an arithmetic sequence are $q - 8$, $-q + 1$, and $2q - 5$. What is the value of the 47th term of this sequence?

    F. 133
    G. 312
    H. 316
    J. 317
    K. 815

15. For some real number q, the first three terms of a geometric sequence are $q - 33$, q, and $q - 66$. What is the fifth term in this sequence?

    A. -176
    B. -88
    C. -44
    D. 88
    E. 176

Sequences
Advanced Problem Set 10

16. For some real number q, the first three terms of a geometric sequence are $6q$, $q+2$, and $2q + \dfrac{4}{9}$. What is the fifth term in this sequence?

    F. $\dfrac{128}{243}$

    G. $\dfrac{2}{3}$

    H. $\dfrac{64}{81}$

    J. $\dfrac{32}{27}$

    K. 2

17. If $\displaystyle\sum_{n=4}^{50}(2n+1) - K = \sum_{n=1}^{46} 2n$, what is the value of k

    A. 46

    B. 92

    C. 144

    D. 230

    E. 423

Sequences
Advanced Problem Set 10

Answer Key

#	Answer	Frequency	Difficulty
1	B	average	1
2	F	average	1
3	D	average	2
4	H	average	1
5	C	average	2
6	J	average	3
7	D	average	4
8	H	average	3
9	B	average	1
10	H	average	1
11	C	average	2
12	H	average	3
13	B	average	2
14	F	average	3
15	A	average	4
16	H	average	4
17	E	average	3

Combinations and Permutations
Advanced Problem Set 11

1. Find the total number of possibilities.
   You are setting the password with 3 letters. You want to use the letters L, M, and P.

2. Find the total number of possibilities.
   A group of 12 people are going to run a race. The top four finishers will advance to the next round.

3. Find the total number of possibilities.
   A team of 17 lacrosse players needs to choose 2 people to meet the ref for a coin toss.

4. Find the total number of possibilities.
   There are 13 students in a class and 4 of them will present on Monday.

DO YOUR FIGURING HERE

Combinations and Permutations
Advanced Problem Set 11

5. Find the total number of possibilities.
   The school service club board had 5 members. They must select a president, vice president, secretary, and treasurer.

6. Find the total number of possibilities.
   A team of 9 basketball players must choose which 5 players are starters.

7. Find the total number of possibilities.
   There are 10 applications for 3 positions: Computer Programmer, Software Tester, and Systems Engineer. (Which position the person gets does not matter.)

8. Find the total number of possibilities.
   A group of 42 runners are going to compete in a race. There will be a gold, silver, and bronze medal.

Combinations and Permutations
Advanced Problem Set 11

9. The five children in the Turkel family are Bo, Clarence, Alejandro, Ronald and Hans. Bo and Ronald are both neither youngest nor oldest. Clarence is the middle child. Bo and Alejandro both have more younger siblings than Ronald. Who is the oldest?

   A. Alejandro
   B. Bo
   C. Clarence
   D. Hans
   E. Ronald

10. How many positive four digit-integers can be made where the first digit must be 9, the last digit cannot be 9 and digits can be repeated?

    F. 90
    G. 900
    H. 999
    J. 1000
    K. 9999

11. Each digit is 3, 5, 7, or 9. Exactly 2 digits are the same. No 2 adjacent digits are the same. How many positive 3-digit integers satisfy all three conditions above?

    A. 3!
    B. 4!
    C. 10
    D. 12
    E. 99

DO YOUR FIGURING HERE

Combinations and Permutations
Advanced Problem Set 11

12. $XOOX$

    The figure above is four bikes that will be assigned to four bikers. Joe and Brian are two of the bikers. What is the probability that both will receive an X bike?

    F. $\dfrac{1}{16}$

    G. $\dfrac{1}{12}$

    H. $\dfrac{1}{9}$

    J. $\dfrac{1}{6}$

    K. $\dfrac{1}{3}$

13. There are 4 roads from Newton to Sparta and 5 roads from Sparta to Stillwater. If Helen drives from Newton to Stillwater and back, passes through Sparta in both directions, and does not travel any road twice, how many different routes for the trip are possible?

    A. 80
    B. 120
    C. 200
    D. 240
    E. 360

14. In a baseball league with 6 teams, each team plays exactly 2 games with each of the other 5 teams in the league. What is the total number of games played in the league?

    F. 6
    G. 18
    H. 30
    J. 32
    K. 36

Combinations and Permutations
Advanced Problem Set 11

15. Any 2 points determine a line. If there are 8 points in a plane, no 3 of which lie on the same line, how many lines are determined by pairs of these 8 points?

    A. 12
    B. 16
    C. 20
    D. 28
    E. 36

16. The Power House Electric company will send a team of 4 electricians to work on a certain job. The company has 5 experienced electricians and 5 trainees. If a team consists of 1 experienced electrician and 3 trainees, how many such teams are possible?

17. A box contains brown plastic poker chips, red plastic poker chips and silver metal poker chips. There are 3 times as many metal poker chips as there are plastic ones. If a chip is randomly picked, the probability that it is red is 4 times the probability that it is brown. If there are 2 brown chips in the box, how many chips are there in total?

    A. 2
    B. 10
    C. 38
    D. 40
    E. 48

Combinations and Permutations
Advanced Problem Set 11

18. A, B, C, D, E, F, G, H

    A list has all possible three-letter arrangements of the letters above with the first letter being H and one of the other letters being F. If no letter is reused in each arrangement and one three-letter arrangement is randomly chosen, then what is the probability of choosing HFD?

    F. $\dfrac{1}{16}$

    G. $\dfrac{1}{12}$

    H. $\dfrac{1}{8}$

    J. $\dfrac{1}{7}$

    K. $\dfrac{1}{6}$

19. A box contains 1 red, 2 white and 3 blue marbles. If 2 marbles are to be drawn randomly, what is the probability that at least one will be white?

    A. $\dfrac{1}{6!}$

    B. $\dfrac{2}{6!}$

    C. $\dfrac{1}{6}$

    D. $\dfrac{1}{3}$

    E. $\dfrac{3}{5}$

Combinations and Permutations
Advanced Problem Set 11

20. In a group of 10 boys and 12 girls, 5 boys and 1 girl speak French. If a boy and a girl are chosen at random, what is the probability that at least one of them speaks French?

    F. $\dfrac{1}{24}$

    G. $\dfrac{1}{12}$

    H. $\dfrac{1}{10}$

    J. $\dfrac{5}{12}$

    K. $\dfrac{13}{24}$

21. The integers 1 through 6 appear on the six faces of a cube, one on each face. If three such cubes are rolled, what is the probability that the sum of the numbers on the top faces is 17 or 18?

    A. $\dfrac{1}{54}$

    B. $\dfrac{1}{48}$

    C. $\dfrac{1}{32}$

    D. $\dfrac{1}{16}$

    E. $\dfrac{1}{12}$

22. A baseball league has 30 teams and each team must play each other team three times. What is the total number of baseball games that are played in this league?

    F. 870

    G. 900

    H. 1305

    J. 1740

    K. 2610

Combinations and Permutations
Advanced Problem Set 11

Answer Key

#	Answer	Frequency	Difficulty
1	6 possibilities	rare	2
2	495 possibilities	rare	2
3	136 possibilities	rare	2
4	715 possibilities	rare	2
5	120 possibilities	rare	2
6	126 possibilities	rare	2
7	120 possibilities	rare	2
8	68,880 possibilities	rare	2
9	A	average	3
10	G	average	3
11	D	average	3
12	J	average	3
13	D	average	3
14	H	average	3
15	D	average	3
16	50	average	3
17	D	average	4
18	G	average	4
19	E	average	4
20	K	average	4
21	A	average	4
22	H	average	4

Imaginary Numbers
Advanced Problem Set 12

1. $i =$

2. $i^2 =$

3. $i^3 =$

4. $i^4 =$

5. $i^5 =$

6. $i^{10} =$

Imaginary Numbers
Advanced Problem Set 12

7. $i^{27} =$

8. $i^{216} =$

9. Considering all complex numbers, where $i^2 = -1$,
$$\frac{3}{1-i}\left(\frac{1+i}{1+i}\right) =$$

   A. $\dfrac{3+i}{1-i}$

   B. $\dfrac{3+i}{1+i}$

   C. $\dfrac{4+i}{2}$

   D. $\dfrac{3+3i}{1-i}$

   E. $\dfrac{3+3i}{2}$

10. Which of the following represents a real, rational number, where $x = -3$

    I. $\sqrt{x^2}$
    II. $\sqrt{x}$
    III. $-\sqrt{(-x)^2}$

    F. I only
    G. II only
    H. I and II only
    J. I and III only
    K. II and III only

Imaginary Numbers
Advanced Problem Set 12

11. Consider the quadratic function, $x^2 + 7x + 49$, which of the following describes the root(s) of the function?

    A. There is one real root
    B. There is one imaginary root
    C. There are two real roots
    D. There are two imaginary roots
    E. There is one real root and two imaginary roots

12. Considering the complex number i and any integer n, which is a possible value for i^n?

    F. -2
    G. -1
    H. 0
    J. 2
    K. 4

13. If x is a positive even integer, which of the following is always an odd integer?

    A. x^4
    B. 4^x
    C. i^4
    D. $4i^4$
    E. $4ix$

14. The following expression $\dfrac{3+4i}{3-4i}$, is equivalent to:

    F. -7
    G. $-7 + 8i$
    H. $\dfrac{24i - 7}{25}$
    J. $\dfrac{25 + 24i}{-7 + 8i}$
    K. $\dfrac{-7 + 8i}{25}$

Imaginary Numbers
Advanced Problem Set 12

Answer Key

#	Answer	Frequency	Difficulty
1	$\sqrt{-1}$ or i	average	2
2	-1	average	2
3	$-\sqrt{-1}$ or $-i$	average	2
4	1	average	2
5	$\sqrt{-1}$ or i	average	2
6	-1	average	2
7	$-\sqrt{-1}$ or $-i$	average	2
8	1	average	2
9	E	average	3
10	J	average	3
11	D	average	4
12	G	average	4
13	C	average	3
14	H	average	4

Matrices
Advanced Problem Set 13

1. $\begin{bmatrix} -6 & 9 \\ 3 & 12 \end{bmatrix} + \begin{bmatrix} 4 & -5 \\ 0 & 1 \end{bmatrix}$

2. $\begin{bmatrix} 5 & 2 \\ 0 & -4 \end{bmatrix} - \begin{bmatrix} -2 & 8 \\ 1 & 3 \end{bmatrix}$

3. $\begin{bmatrix} 8 & -6 & 3 \\ 2 & 4 & -1 \\ 7 & 0 & 1 \end{bmatrix} + \begin{bmatrix} 3 & 7 & -2 \\ -9 & 1 & 4 \\ -6 & 2 & 1 \end{bmatrix}$

4. $\begin{bmatrix} 3 & -1 & 8 \\ 1 & 2 & 4 \\ 0 & 3 & 6 \end{bmatrix} - \begin{bmatrix} 2 & -4 & 5 \\ 0 & 3 & 8 \\ -9 & 7 & 1 \end{bmatrix}$

5. What are the dimensions of the resulting matrix?
$\begin{bmatrix} 1 & 8 & 0 \\ 6 & 3 & 5 \\ 2 & -4 & -7 \end{bmatrix} \times \begin{bmatrix} 4 & 2 & 7 & -3 & -6 & 0 \\ 6 & 5 & 8 & 4 & 3 & 1 \\ 1 & -4 & -2 & 8 & -6 & -7 \end{bmatrix}$

6. What are the dimensions of the resulting matrix?
$\begin{bmatrix} 1 & 8 \\ -3 & 2 \\ 8 & -4 \\ 0 & 7 \end{bmatrix} \times \begin{bmatrix} 4 & 2 & 7 & -3 & -6 & 0 \\ 6 & 5 & 8 & 4 & 3 & 1 \end{bmatrix}$

7. Give the determinant for the following:
$\begin{bmatrix} 1 & 8 \\ 4 & 16 \end{bmatrix}$

8. Give the determinant for the following:
$\begin{bmatrix} 7 & 6 \\ -2 & -3 \end{bmatrix}$

9. Give the determinant for the following:
$\begin{bmatrix} 9 & 10 \\ 10 & 11 \end{bmatrix}$

10. Give the determinant for the following:
$\begin{bmatrix} 19 & 1 \\ 26 & 4 \end{bmatrix}$

DO YOUR FIGURING HERE

Matrices
Advanced Problem Set 13

11. Give the determinant for the following:
$$\begin{bmatrix} 1 & 2 \\ 3 & 4 \end{bmatrix}$$

12. Give the determinant for the following:
$$\begin{bmatrix} -3 & -8 \\ 1 & 4 \end{bmatrix}$$

13. Give the determinant for the following:
$$\begin{bmatrix} 1 & 3 & 9 \\ 0 & 4 & 6 \\ -2 & -1 & 1 \end{bmatrix}$$

14. Give the determinant for the following:
$$\begin{bmatrix} 3 & -1 & 0 \\ 0 & 4 & 2 \\ -2 & -3 & 1 \end{bmatrix}$$

15. Give the determinant for the following:
$$\begin{bmatrix} 0 & 4 & 2 \\ -6 & 3 & 7 \\ 1 & 0 & 9 \end{bmatrix}$$

16. Give the determinant for the following:
$$\begin{bmatrix} 3 & 9 & 1 \\ -4 & 2 & 8 \\ 7 & 5 & -2 \end{bmatrix}$$

17. Which of the following matrices satisfies the following equation?
$$\begin{bmatrix} ? & ? \\ ? & ? \end{bmatrix} \begin{bmatrix} 1 \\ 0 \end{bmatrix} = \begin{bmatrix} 1 \\ -1 \end{bmatrix}$$

A. $\begin{bmatrix} 1 & 1 \\ -1 & -1 \end{bmatrix}$

B. $\begin{bmatrix} -1 & -1 \\ 1 & -1 \end{bmatrix}$

C. $\begin{bmatrix} -1 & 1 \\ 1 & -1 \end{bmatrix}$

D. $\begin{bmatrix} 1 & -1 \\ 1 & -1 \end{bmatrix}$

E. $\begin{bmatrix} 1 & 1 \\ 1 & 1 \end{bmatrix}$

DO YOUR FIGURING HERE

Matrices
Advanced Problem Set 13

18. Which of the following matrices satisfies the following equation?

    $$\begin{bmatrix} ? & ? \\ ? & ? \end{bmatrix} \begin{bmatrix} 1 \\ 2 \end{bmatrix} = \begin{bmatrix} 1 \\ 3 \end{bmatrix}$$

    F. $\begin{bmatrix} 1 & 2 \\ 1 & 3 \end{bmatrix}$

    G. $\begin{bmatrix} 3 & -1 \\ -1 & 2 \end{bmatrix}$

    H. $\begin{bmatrix} 3 & 1 \\ 1 & 2 \end{bmatrix}$

    J. $\begin{bmatrix} -1 & 3 \\ 2 & -1 \end{bmatrix}$

    K. $\begin{bmatrix} -1 & 2 \\ 3 & -1 \end{bmatrix}$

19. Which of the following matrices satisfies the following equation?

    $$\begin{bmatrix} ? & ? \\ ? & ? \end{bmatrix} \begin{bmatrix} -1 \\ 3 \end{bmatrix} = \begin{bmatrix} 7 \\ -7 \end{bmatrix}$$

    A. $\begin{bmatrix} -2 & 3 \\ -4 & 1 \end{bmatrix}$

    B. $\begin{bmatrix} 4 & 3 \\ 2 & -1 \end{bmatrix}$

    C. $\begin{bmatrix} 2 & 3 \\ 4 & -1 \end{bmatrix}$

    D. $\begin{bmatrix} -1 & 2 \\ 3 & 4 \end{bmatrix}$

    E. $\begin{bmatrix} 4 & -1 \\ 2 & 3 \end{bmatrix}$

DO YOUR FIGURING HERE

Matrices
Advanced Problem Set 13

20. Which of the following matrices satisfies the following equation?

$$\begin{bmatrix} -5 & 5 \\ 5 & -5 \end{bmatrix} \begin{bmatrix} ? \\ ? \end{bmatrix} = \begin{bmatrix} 10 \\ -10 \end{bmatrix}$$

F. $\begin{bmatrix} -1 \\ -1 \end{bmatrix}$

G. $\begin{bmatrix} 1 \\ 1 \end{bmatrix}$

H. $\begin{bmatrix} 0 \\ 0 \end{bmatrix}$

J. $\begin{bmatrix} 1 \\ -1 \end{bmatrix}$

K. $\begin{bmatrix} -1 \\ 1 \end{bmatrix}$

21. Which of the following matrices satisfies the following equation?

$$\begin{bmatrix} 1 & -1 \\ -1 & 1 \end{bmatrix} \begin{bmatrix} ? \\ ? \end{bmatrix} = \begin{bmatrix} -2 \\ 2 \end{bmatrix}$$

A. $\begin{bmatrix} 5 \\ 3 \end{bmatrix}$

B. $\begin{bmatrix} 4 \\ 2 \end{bmatrix}$

C. $\begin{bmatrix} 2 \\ 5 \end{bmatrix}$

D. $\begin{bmatrix} 3 \\ 5 \end{bmatrix}$

E. $\begin{bmatrix} 4 \\ 3 \end{bmatrix}$

DO YOUR FIGURING HERE

Matrices
Advanced Problem Set 13

22. The 2 x 3 matrix $\begin{bmatrix} -4 & 0 & 3 \\ 2 & 5 & 1 \end{bmatrix}$ represents $\triangle ABC$ with vertices $A(-4, 2)$, $B(0, 5)$, and $C(3, 1)$ in the standard xy-coordinate plane. After a translation of $\triangle ABC$, the matrix representing the new triangle is $\begin{bmatrix} -2 & 2 & 5 \\ 1 & 4 & n \end{bmatrix}$. What is the value of n?

   F. -2
   G. -1
   H. 0
   J. 1
   K. 2

23. Given that
    $2\begin{bmatrix} 4 & -1 \\ x & 3 \end{bmatrix} + \begin{bmatrix} 4 & 4 \\ 2 & 1 \end{bmatrix} = \begin{bmatrix} 12 & 2 \\ 10 & 7 \end{bmatrix}$
    what is the value of x?

    A. 3.5
    B. 4
    C. 4.5
    D. 5.5
    E. 8

Matrices
Advanced Problem Set 13

24. The number of students participating in certain winter sports at a local high shcool is given by the following matrix:

$[\,6 \quad 30 \quad 40 \quad 80\,]$

The athletic director estimates the ratio of the number of students who earn varsity letters to the number of total students playing the sport with the matrix below. Given these matrices, what is the athletic director's estimate for the number of varsity letters given out in these sports?

Skiing $\begin{bmatrix} 0.5 \\ 0.7 \\ 0.3 \\ 0.8 \end{bmatrix}$
Basketball
Hockey
Track & Field

- **F.** 93
- **G.** 198
- **H.** 107
- **J.** 112
- **K.** 104

25. The 2 x 3 matrix $\begin{bmatrix} -6 & 1 & 4 \\ 3 & 7 & 2 \end{bmatrix}$ represents $\triangle ABC$ with vertices $A(-6, 3)$, $B(1, 7)$, and $C(4, 2)$ in the standard xy-coordinate plane. After a translation of $\triangle ABC$, the matrix representing the new triangle is $\begin{bmatrix} -1 & 6 & 9 \\ 4 & 8 & n \end{bmatrix}$. What is the value of n?

- **A.** 1
- **B.** 2
- **C.** 3
- **D.** 4
- **E.** 5

Matrices
Advanced Problem Set 13

26. The determinant of the matrix
$\begin{bmatrix} e & f \\ g & h \end{bmatrix}$ is $eh - fg$.

    The determinant of the following array is 0.
    $\begin{bmatrix} (x-4) & 11 \\ 4 & (x+3) \end{bmatrix}$

    What are the possible values of x?

    F. -8 and 7

    G. -7 and 8

    H. -4 and 3

    J. -3 and 4

    K. $\dfrac{1}{2} - \dfrac{\sqrt{124}}{2}$ and $\dfrac{1}{2} + \dfrac{\sqrt{124}}{2}$

27. Given that
    $c \begin{bmatrix} 6 & 8 \\ 3 & 4 \end{bmatrix} = \begin{bmatrix} d & 52 \\ e & f \end{bmatrix}$

    for some real number c, what is $d + f$.

    A. $\dfrac{13}{2}$

    B. 52

    C. 26

    D. 39

    E. 65

Matrices
Advanced Problem Set 13

Answer Key

#	Answer	Frequency	Difficulty
1	$\begin{bmatrix} -2 & 4 \\ 3 & 13 \end{bmatrix}$	average	3
2	$\begin{bmatrix} 7 & -6 \\ -1 & -7 \end{bmatrix}$	average	3
3	$\begin{bmatrix} 11 & 1 & 1 \\ -7 & 5 & 3 \\ 1 & 2 & 2 \end{bmatrix}$	average	3
4	$\begin{bmatrix} 1 & 3 & 3 \\ 1 & -1 & -4 \\ 9 & -4 & 5 \end{bmatrix}$	average	3
5	3X6	average	3
6	4X6	average	3
7	-16	average	3
8	-9	average	3
9	-1	average	3
10	50	average	3
11	-2	average	3
12	-4	average	3
13	46	average	3
14	34	average	3
15	238	average	3
16	266	average	3
17	A	average	3
18	G	average	3
19	C	average	3
20	K	average	3
21	D	average	3
22	H	average	3
23	B	average	3
24	H	average	3
25	C	average	3
26	G	average	3
27	E	average	3

Advanced Mathematics Mixed Problem Set 1
Advanced Problem Set 14

1. If a store owner marks up merchandise 90% and sells a TV for $950, how much did the owner pay for the TV?

2. The gorilla population in Africa decreases by 20% each year. If there are 4000 gorillas in 1999, then how many will there be in 2012? Round to the nearest whole number.

3. The ratio of dogs to cats at a pet store is 5 to 3. If there are 96 animals total, then how many more dogs than cats are there?

4. Jesse makes 25 calls on his first day as a telemarketer and 6 more calls each day than the previous day. How many total calls did Jesse make during his first 48 days at work?

5. Demitch's Loveline charges $9 total for the initial 4 minutes and $4 for each additional minute. Miness's Hotline charges $20 total for the initial 3 minutes and $1 for each additional minute. At what minute will both calls cost the same amount?

6. The pattern: *tick, tock, clack, clock, click* repeats for 8,342 words. What is the last word?

7. There are 4 roads from Chappaqua to Armonk and 5 roads from Armonk to Mt. Kisco. How many routes can you drive from Chappaqua to Mt. Kisco through Armonk if you take each road only once?

DO YOUR FIGURING HERE

Advanced Mathematics Mixed Problem Set 1
Advanced Problem Set 14

8.   ABCDEABCDEABCD...
If the sequence above starts on A and contains 364 letters, then what is the last letter of the sequence?

9. Every 12th customer at a store receives a 20% off coupon, and every 15th customer receives a t-shirt. If there were 980 customers, then what is the probability of a customer getting both a coupon and a t-shirt?

10. How many starting lineups of 5 basketball players are possible on a team of 10 if only 2 of the players can play power forward and all of the players can play all other four positions?

11. A circle has its center at $(6, 4)$ and is tangent to the x-axis. Another circle has the same center but is tangent to the y-axis. What is the difference in area between the 2 circles?

12. $g(x) = x^2 - 1$
$t(x) = g(4x) + 5$
$t(a) = 260$
What is the value of $g(a)$?

Advanced Mathematics Mixed Problem Set 1
Advanced Problem Set 14

13. $g(x) = 5x + 3$
    $p(x) = g(4x) + 8$
    $g(b) = 198$
    What is the value of $p(b)$?

14. $\dfrac{n!}{(n-1)!} = 5$
    What is the value of $(n+1)!$?

15. $g(x) = x^2 + 1$
    $f(x) = g(3x) - 4$
    $f(a) = 897$
    What is the value of a?

16. $f(x) = x^2 + 9$
    $g(x) = f(2x) - 5$
    $g(a) = 488$
    What is the value of a?

Advanced Mathematics Mixed Problem Set 1
Advanced Problem Set 14

17. $h(x) = 2x - 1$
$k(x) = h\left(\frac{x}{2}\right) + 1$
$k(b) = 10$
What is the value of b?

18. The area of square $EACD$ is 36. It is inscribed in a circle with the center D. What is the area of the shaded region in terms of π?

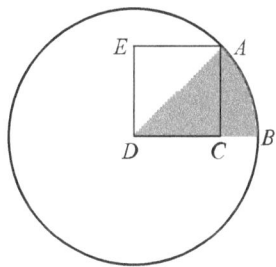

19. If $ABCD$ is a square with an area of 100, then what is the circumference of the inscribed circle in terms of π?

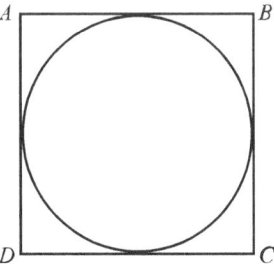

Advanced Mathematics Mixed Problem Set 1
Advanced Problem Set 14

20. Rhombus $ABCD$ has a perimeter of 24. A and C are centers of their respective congruent circles. What is the length of the darkened arcs, in terms of π?

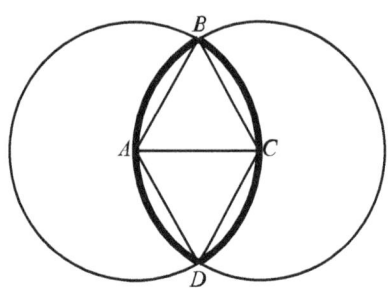

DO YOUR FIGURING HERE

21. What is the area of an equilateral triangle that has a perimeter of 120? Please give the answer in radical form.

22. What is the value of $b - a$ in the regular octagon below?

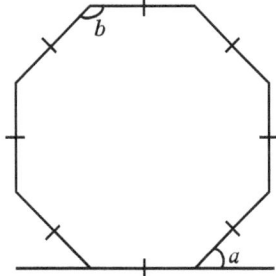

23. What is the equation of the line $x + 4y = 2$ after it has been reflected over the x-axis? Please give your answer in slope-intercept form.

24. 350 students attended a high school dance. If the girl to boy ratio is 18 to 17, then how many more girls than boys attended?

Advanced Mathematics Mixed Problem Set 1
Advanced Problem Set 14

25. If line segment $\overline{BD}$ (not shown) has a length of 40, what is the area of the circle?

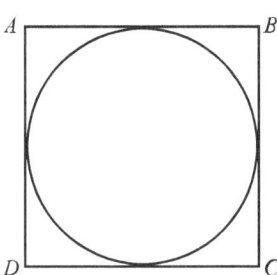

26. $ABCD$ is a trapezoid with an area of 140. If the length of $\overline{AB}$ is 5 and the length of $\overline{AE}$ is 7, then what is the length of $\overline{CD}$?

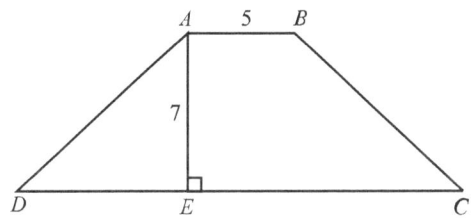

27. If the area of equilateral triangle ABC is $256\sqrt{3}$ then what is the surface area of the cube?

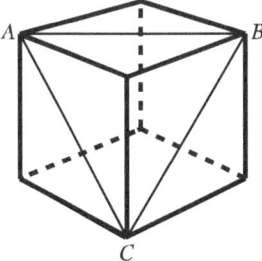

DO YOUR FIGURING HERE

Advanced Mathematics Mixed Problem Set 1
Advanced Problem Set 14

28. If ABC is an equilateral triangle with perimeter of 90 and the height of the cylinder is 15, then what is the volume of the cylinder?

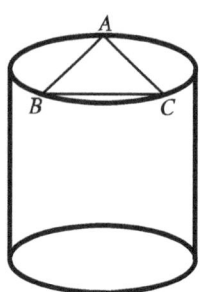

DO YOUR FIGURING HERE

Advanced Mathematics Mixed Problem Set 1
Advanced Problem Set 14

Answer Key

#	Answer	Frequency	Difficulty
1	500	popular	2
2	220	popular	3
3	24	popular	3
4	7,968	average	4
5	8	average	3
6	*tock*	average	2
7	20	rare	4
8	D	average	3
9	$\frac{1}{196}$	average	3
10	6048	rare	4
11	20π	popular	3
12	15	popular	4
13	791	popular	4
14	720	rare	3
15	± 10	popular	3
16	± 11	popular	3
17	10	popular	3
18	9π	popular	4
19	10π	popular	3
20	8π	popular	5
21	$400\sqrt{3}$ OR 692.82	popular	4
22	90	popular	3
23	$y = \frac{1}{4}x - \frac{1}{2}$	average	3
24	10	popular	2
25	200π OR 628.3	popular	2
26	35	popular	2
27	3,072	popular	5
28	$4,500\pi$	popular	5

Advanced Mathematics Mixed Problem Set 2
Advanced Problem Set 15

1. $\dfrac{6}{\sqrt{5}} + \dfrac{3}{\sqrt{7}} = ?$

   A. $\dfrac{6\sqrt{7} + 3\sqrt{5}}{\sqrt{12}}$

   B. $\dfrac{6\sqrt{7} + 3\sqrt{5}}{\sqrt{35}}$

   C. $\dfrac{9}{\sqrt{5} + \sqrt{7}}$

   D. $\dfrac{9}{\sqrt{12}}$

   E. $\dfrac{9}{\sqrt{35}}$

DO YOUR FIGURING HERE

2. The system of equations below has multiple solutions, all of which satisfy the equation $y = \dfrac{5}{3}x - 4$. If it can be determined, what is the value of c?

$$10x - 6y = 24$$
$$25x - cy = 60$$

   F. -6

   G. 4

   H. 5

   J. 15

   K. Cannot be determined

3. Which of the following gives the fractions $-\dfrac{4}{7}, -\dfrac{7}{9}$ and $-\dfrac{3}{4}$ in order from the least to greatest?

   A. $-\dfrac{7}{9} < -\dfrac{3}{4} < -\dfrac{4}{7}$

   B. $-\dfrac{3}{4} < -\dfrac{4}{7} < -\dfrac{7}{9}$

   C. $-\dfrac{7}{9} < -\dfrac{4}{7} < -\dfrac{3}{4}$

   D. $-\dfrac{3}{4} < -\dfrac{7}{9} < -\dfrac{4}{7}$

   E. $-\dfrac{4}{7} < -\dfrac{3}{4} < -\dfrac{7}{9}$

Advanced Mathematics Mixed Problem Set 2
Advanced Problem Set 15

4. Consider the 2 functions $f(x) = 3x + 6$ and $g(x) = 4x + c$ where c is a real number. If $f(g(x)) = g(f(x))$, then $c = ?$

    F. 0

    G. -3

    H. 9

    J. 12

    K. Any real number

5. $\dfrac{\frac{5}{x+4}}{1 - \frac{8}{x+4}} = ?$

    A. $\dfrac{5}{x-4}$

    B. -4

    C. 4

    D. $\dfrac{5}{x^2 + 16}$

    E. $\dfrac{5}{x+4}$

6. The product of 2 numbers is 41. If 1 of the numbers is the complex number $5 + 4i$, what is the other?

    F. $41 - 5i$

    G. $41i$

    H. $\dfrac{4}{25} + \dfrac{1}{5}i$

    J. $5 - 4i$

    K. $25 + 5i$

Advanced Mathematics Mixed Problem Set 2
Advanced Problem Set 15

7. Miles and Christian each ran 8 laps around a 600 meter track without stopping. Miles ran at a constant speed of 240 meters per minute. Christian ran at a constant speed of 180 meters per minute. Both Miles and Christian began running at the same instant. How many laps did Christian have left to run when Miles had completed his run?

   A. 0.2
   B. 1.0
   C. 1.3
   D. 2
   E. 6.7

8. Larry rode his bike, mostly downhill, to visit Kyle. The trip to Kyle's house took t minutes. Returning home, mostly uphill, Larry was able to travel at an average speed one half that of his trip to Kyle's. Which of the following is an expression for the total number of minutes Larry bicycled on the entire trip?

   F. $\dfrac{3t}{2}$ minutes
   G. t minutes
   H. $t + 2$ minutes
   J. $2t$ minutes
   K. $3t$ minutes

9. Which of the following is a quadratic equation that has $-\dfrac{3}{4}$ as its only solution?

   A. $16x^2 - 24x + 9 = 0$
   B. $16x^2 + 24x + 9 = 0$
   C. $16x^2 + 24x - 9 = 0$
   D. $16x^2 + 9 = 0$
   E. $16x^2 - 9 = 0$

Advanced Mathematics Mixed Problem Set 2
Advanced Problem Set 15

10. One hot-air balloon is 14 km east and 6 km north of an airport tower while a second hot-air balloon at the same altitude is 4 km west and 9 km south of the same airport tower. Approximately how many kilometers separate the 2 hot-air balloons?

    F. 9.84 km

    G. 15.2 km

    H. 18 km

    J. 23.4 km

    K. 25 km

11. Which of the following binomials is a factor of $5x^2 + 12x - 9$?

    A. $x - 3$

    B. $x - 4$

    C. $x + 3$

    D. $x + 4$

    E. $x + 2$

12. If $0° < x \leq 90°$, and $2\cos^2 x - 1 = 0$, then $x =$

    F. $0°$

    G. $30°$

    H. $45°$

    J. $60°$

    K. $90°$

13. Let θ be the radian angle measure that satisfies $\cos^2\theta - \cos\theta = -\frac{1}{4}$ for $0 < \theta < \frac{\pi}{2}$. What is $\sin\theta$?

    A. $\frac{1}{4}$

    B. $\frac{1}{8}$

    C. $\frac{1}{18}$

    D. $\frac{\sqrt{3}}{2}$

    E. $\frac{1}{2}$

DO YOUR FIGURING HERE

Advanced Mathematics Mixed Problem Set 2
Advanced Problem Set 15

14. Given that $5\cos(a) = 0$ and $3\sin(\pi - b) = 3$, which of the following could be a value, in radians, of $a + b$?

    F. 0

    G. $\dfrac{\pi}{2}$

    H. 2

    J. $\dfrac{3\pi}{2}$

    K. π

15. Given that $\sin A = \dfrac{7}{25}$ and $0° \leq A° < 360°$, what are all possible values of $\cos A$?

    A. $-\dfrac{7}{25}$ only

    B. $-\dfrac{7}{25}$ and $\dfrac{7}{25}$

    C. $\dfrac{24}{25}$ only

    D. $-\dfrac{24}{25}$ only

    E. $-\dfrac{24}{25}$ and $\dfrac{24}{25}$

16. Amy mixes 70 mL of Solution C with 30 mL of Solution Y. Solution C has a 60% of hydrobromic acid concentration; Solution Y has an unknown hydrobromic acid concentration. When Amy tests the resulting 100 mL solution, she finds that it has a 54% hydrobromic acid concentration. What is the approximate hydrobromic acid concentration of Solution Y?

    F. 20%

    G. 24%

    H. 34%

    J. 40%

    K. 51%

DO YOUR FIGURING HERE

Advanced Mathematics Mixed Problem Set 2
Advanced Problem Set 15

17. Each of 30 students in a class took a test and received a whole number score. The median of the the scores was 85. None of the students received a score of 85, and 20% received scores of 90 or above. How many students received scores of 86 to 89?

    A. 4
    B. 8
    C. 9
    D. 10
    E. 12

18. 10 cars containing a total of 30 people crossed a toll bridge. Each of the 10 cars contained at least 1 person but no more than 5 people. At most how many cars contained exactly 4 people?

    F. 5
    G. 6
    H. 7
    J. 8
    K. 9

19. The quantity $\sqrt[m]{2^h}$ is defined when m is an integer greater than 2 and h is any nonzero real number. Which of the following is a relationship between m and h that will always make $\sqrt[m]{2^h}$ a positive integer?

    A. $\dfrac{h}{m}$ is a positive integer
    B. $\dfrac{m}{h}$ is a positive integer
    C. m is greater than h
    D. h is greater than m
    E. The sum of h and m is 1

Advanced Mathematics Mixed Problem Set 2
Advanced Problem Set 15

20. The Recreation Department wants to build a circular wading pool in the city park. The area available for the pool is a fenced-in rectangular region 16 meters by 20 meters. If the recreation department wants the wading pool to be as large as possible, and the edge of the pool must be at least 2 meters from the fence all around, how many meters long should the radius of the pool be?

    F. 5
    G. 6
    H. 8
    J. 10
    K. 12

Advanced Mathematics Mixed Problem Set 2
Advanced Problem Set 15

Answer Key

#	Answer	Frequency	Difficulty
1	B	popular	2
2	J	popular	3
3	A	popular	2
4	H	popular	3
5	A	popular	3
6	J	average	4
7	D	rare	4
8	K	popular	3
9	B	popular	3
10	J	popular	3
11	C	popular	3
12	H	popular	4
13	D	popular	4
14	K	popular	5
15	E	popular	5
16	J	popular	4
17	C	popular	3
18	G	popular	3
19	A	popular	3
20	G	popular	2

Advanced Mathematics Mixed Problem Set 3
Advanced Problem Set 16

1. Frederick is loading a truck using a ramp, as shown below. The ramp is 13 feet long, and the end of the ramp that is resting on the truck is 1.7 feet above the level ground and the other end is touching the ground. Which of the following expressions gives the angle of elevation of the ramp?

   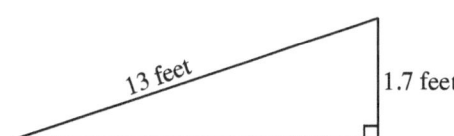

   A. $\arctan\left(\dfrac{1.7}{13}\right)$

   B. $\arccos\left(\dfrac{1.7}{13}\right)$

   C. $\arcsin\left(\dfrac{13}{1.7}\right)$

   D. $\arccos\left(\dfrac{13}{1.7}\right)$

   E. $\arcsin\left(\dfrac{1.7}{13}\right)$

2. To make a 1000-piece jigsaw puzzle more challenging, a puzzle company includes 10 extra pieces in the box along with the 1000 pieces. Those 10 extra pieces do not fit anywhere in the puzzle. If you buy such a puzzle box, break the seal on the box, and immediately select 1 piece at random, what is the probability that it will be 1 of the extra pieces?

   F. $\dfrac{1}{10}$

   G. $\dfrac{1}{1000}$

   H. $\dfrac{1}{100}$

   J. $\dfrac{1}{1010}$

   K. $\dfrac{1}{101}$

Advanced Mathematics Mixed Problem Set 3
Advanced Problem Set 16

3. In the figure below, lines g and h are parallel. Lines i and j intersect g at the same point.

   Angle $a = 4x + 12$
   Angle $b = 3x - 18$
   Angle $c = 5x - 18$

   What is the value of x?

   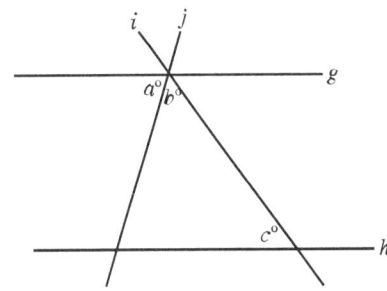

   A. 10
   B. 13
   C. 16
   D. 17
   E. 18

4. Jenna and Matt each ran 8 laps around a 600-meter track without stopping. Jenna ran at a constant speed of 240 meters per minute. Matt ran at a constant speed of 180 meters per minute. Both Jenna and Matt began running at the same instant. How many laps did Matt have left to run when Jenna had completed her run?

   F. $\dfrac{1}{2}$
   G. $\dfrac{6}{5}$
   H. 1
   J. 2
   K. $\dfrac{8}{3}$

Advanced Mathematics Mixed Problem Set 3
Advanced Problem Set 16

5. For what value of c would the following system of equations have an infinite number of solutions?
$$3x + 4y = 16$$
$$12x + 16y = 8c$$

    A. 2
    B. 4
    C. 8
    D. 16
    E. 64

6. Given that $4\cos a = 4$ and $4\sin(\pi - b) = 4$, which of the following could be a value, in radians, of $a + b$?

    F. 0
    G. $\dfrac{\pi}{2}$
    H. π
    J. $\dfrac{3\pi}{2}$
    K. 2

7. One of the following equations determines the graph in the standard xy-coordinate plane below. Which one?

    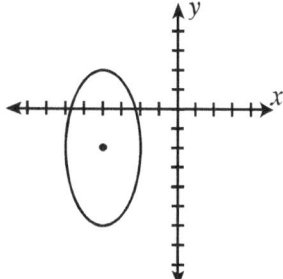

    A. $\dfrac{(x+4)^2}{4} + \dfrac{(y+2)^2}{16} = 1$

    B. $\dfrac{(x-4)^2}{16} + \dfrac{(y-2)^2}{4} = 1$

    C. $\dfrac{(x+2)^2}{4} + \dfrac{(y+4)^2}{16} = 1$

    D. $\dfrac{(x-4)^2}{4} + \dfrac{(y-2)^2}{16} = 1$

    E. $\dfrac{(x-4)^2}{4} + \dfrac{(y+2)^2}{16} = 1$

Advanced Mathematics Mixed Problem Set 3
Advanced Problem Set 16

8. Cubes each having a side length of 1 cm are put together to form a rectangular solid with 9 layers. Each layer has 4 cubes across and is 1 cube deep. What is the volume, in cubic centimeters, of the rectangular solid?

   F. 9
   G. 18
   H. 24
   J. 36
   K. 48

9. Consider the 2 functions $g(x) = 5x + 8$ and $h(x) = 2x + c$, where c is a real number. If $g(h(x)) = h(g(x))$, then $c = ?$

   A. 0
   B. 2
   C. 4
   D. 8
   E. All Real Numbers

10. Steve and Kyrie are saving to make a down payment on a house. With an initial deposit of 17,000 dollars, they have opened an account that compounds interest at an annual rate of 4.7%. Assuming that Steve and Kyrie make no additional deposits or withdrawals, which of the following expressions gives the dollar value of the account 8 years after the initial deposit?
    Note: For an account with an initial deposit of I dollars that compounds interest at an annual rate of $r\%$. The value of the account y years after the initial deposit is $I\left(1 + \dfrac{r}{100}\right)^y$ dollars.

    F. $17000 \left(1.47\right)^8$
    G. $17000 \left(1.047\right)^8$
    H. $17000 \left(5.7\right)^8$
    J. $17000 \left(4.7\right)^8$
    K. $17000^8 \left(1.47\right)$

DO YOUR FIGURING HERE

Advanced Mathematics Mixed Problem Set 3
Advanced Problem Set 16

11. The radius of the base of the right circular cone shown below is 3 inches, and the height of the cone is 8 inches. Solving which of the following equations gives the measure, θ, of the angle formed by a slant height of the cone and a radius?

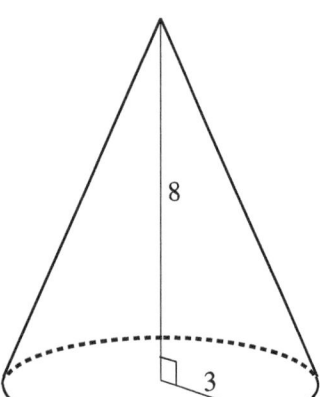

A. $\tan(\theta) = \dfrac{8}{3}$

B. $\tan(\theta) = \dfrac{3}{8}$

C. $\cos(\theta) = \dfrac{8}{3}$

D. $\cos(\theta) = \dfrac{3}{8}$

E. $\sin(\theta) = \dfrac{8}{3}$

Advanced Mathematics Mixed Problem Set 3
Advanced Problem Set 16

12. The solid shown below is composed of a right circular cylinder and a right circular cone with base diameters and heights given in centimeters. The cylinder and the cone have equal base diameters. What is the volume, in cubic centimeters, of the solid?

    Note: The volume of a right circular cylinder with a base radius r and a height h is $\pi r^2 h$. The volume of a right circular cone with base radius r and height h is $\frac{\pi r^2 h}{3}$.

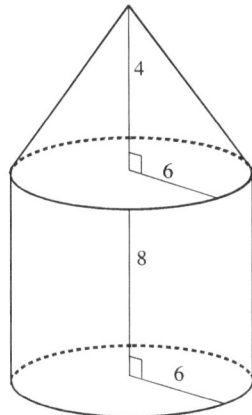

F. 288π

G. 336π

H. 1344π

J. 48π

K. 432π

Advanced Mathematics Mixed Problem Set 3
Advanced Problem Set 16

13. Use the figure below to answer questions 13 through 15.
    Trapezoid $EFGH$ is graphed in the standard xy-coordinate plane below. What is the slope of $\overline{FE}$?

    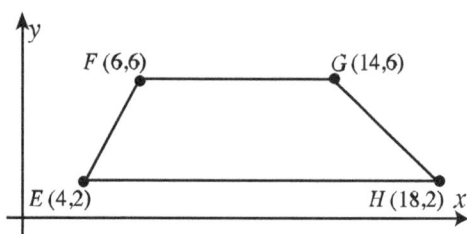

    A. -2
    B. $\frac{1}{2}$
    C. 2
    D. $-\frac{1}{2}$
    E. 4

14. Refer to the figure in question 13.
    When trapezoid $EFGH$, below, is reflected over the x-axis, what are the coordinates of G?

    F. $(14, -6)$
    G. $(-14, 6)$
    H. $(6, -14)$
    J. $(-14, -6)$
    K. $(-6, 14)$

15. Refer to the figure in question 13.
    Which of the following vertical lines cuts $EFGH$, shown below, into 2 trapezoids with equal areas?

    A. $x = 9.5$
    B. $x = 10.5$
    C. $x = 11.5$
    D. $x = 12.5$
    E. $x = 13.5$

Advanced Mathematics Mixed Problem Set 3
Advanced Problem Set 16

16. $\dfrac{8}{\sqrt{5}} + \dfrac{7}{\sqrt{3}} =$

    F. $\dfrac{8\sqrt{3} + 7\sqrt{5}}{\sqrt{8}}$

    G. $\dfrac{8\sqrt{3} + 7\sqrt{5}}{\sqrt{15}}$

    H. $\dfrac{15}{\sqrt{15}}$

    J. $\dfrac{15}{\sqrt{8}}$

    K. $\dfrac{15}{\sqrt{5} + \sqrt{3}}$

17. Charlie mixes 70 mL of solution C with 30 mL of solution Y. Solution C has a 60% HCl concentration; solution Y has an unknown HCl concentration. When Charlie tests the resulting 100 mL solution, he finds that it has a 54% HCl concentration. What is the HCl concentration of solution Y?

    A. 20%

    B. 24%

    C. 34%

    D. 40%

    E. 50%

18. What are the real number values of x that make the equation $\sqrt[4]{x^{24}} = x^6$ true?

    F. All Real Numbers

    G. $x < 0$

    H. $x > 0$

    J. $x \leq 0$

    K. $x \geq 0$

Advanced Mathematics Mixed Problem Set 3
Advanced Problem Set 16

19. Use the figure below to answer questions 19 through 21.

    At West High School, the 22 cast members of the fall show sold only adult and student tickets for both Friday and Saturday night performances. The price of each student ticket was $4 and the price of each adult ticket was $6. The table below gives the number of tickets sold, by type and by night. The stem and leaf plot below shows the number of tickets, regardless of type, sold by each of the 22 cast members.

    Suppose 1 cast member will be picked at random from the 22 cast members who sold tickets to receive a prize. What is the probability of selecting a cast member who has sold more than 18 tickets?

Ticket Type	Friday	Saturday
Student	116	104
Adult	84	96

Stem	Leaf
0	1 8 9
1	1 3 4 4 4 6 6 8 8 9
2	0 0 2 4 5 7 7
3	1 3

    A. $\dfrac{1}{4}$

    B. $\dfrac{5}{11}$

    C. $\dfrac{6}{11}$

    D. $\dfrac{10}{11}$

    E. $\dfrac{4}{5}$

20. Refer to the information given in question 19. For which night was the total amount collected for the tickets greater, and by how much?

    F. Friday by $24

    G. Friday by $32

    H. Friday by $64

    J. Saturday by $24

    K. Saturday by $32

Advanced Mathematics Mixed Problem Set 3
Advanced Problem Set 16

21. Refer to the information given in question 19. The auditorium where the fall show will be performed has 12 seats in the front row. Each row behind the first row has 6 more seats than the row in front of it.
    How many seats are in the 7th row?

    A. 42
    B. 44
    C. 48
    D. 54
    E. 60

22. The shaded region in the graph below represents the solution set to which of the following sets of inequalities?

    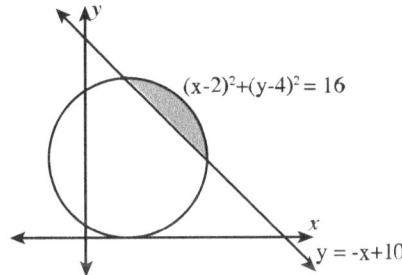

    F. $y > -x + 10$ and
       $(x - 2)^2 + (y - 4)^2 > 16$

    G. $y < -x + 10$ and
       $(x - 2)^2 + (y - 4)^2 > 16$

    H. $y < -x + 10$ and
       $(x - 2)^2 + (y - 4)^2 < 16$

    J. $y > -x + 10$ and
       $(x - 2)^2 + (y - 4)^2 < 16$

    K. $y - 4 < 4$ and $x - 2 > 2$

23. Which of the following solution sets shows the answer to the inequality $|x - 7| > -4$?

    A. $(3, 11)$
    B. $(11, \infty)$
    C. $(-\infty, 3)$
    D. $(-\infty, \infty)$
    E. $(\emptyset)$

Advanced Mathematics Mixed Problem Set 3
Advanced Problem Set 16

24. The sides of an acute triangle measure 11m, 17m and 20m. Which of the following equations when solved for θ gives the measure of the smallest angle of the triangle?

Note: for any triangle with sides of length a, b, and c that are opposite angles A, B and C, respectively, $\dfrac{\sin A}{a} = \dfrac{\sin B}{b} = \dfrac{\sin C}{c}$ and $c^2 = a^2 + b^2 - 2ab\cos C$.

F. $\dfrac{\sin \theta}{11} = \dfrac{1}{17}$

G. $\dfrac{\sin \theta}{11} = \dfrac{1}{20}$

H. $11^2 = 17^2 + 20^2 - 2(17)(20)\cos \theta$

J. $17^2 = 11^2 + 20^2 - 2(11)(20)\cos \theta$

K. $20^2 = 17^2 + 11^2 - 2(17)(11)\cos \theta$

DO YOUR FIGURING HERE

Advanced Mathematics Mixed Problem Set 3
Advanced Problem Set 16

25. Engineers are building a straight underwater pipeline from point A, which is located on an island, to point B, which is located on a straight stretch of coastline, as shown on the map below. The length of the pipeline is 13.7 miles. Point C is located 4.8 miles along the coastline from point B and the line $\overline{AC}$ forms a right angle with the coastline. Which of the following expressions gives the measure of the acute angle formed by the pipeline and the coastline?

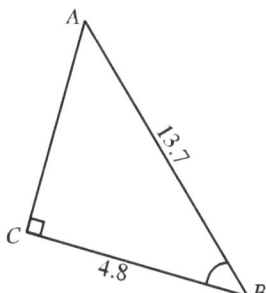

A. $\arcsin\left(\dfrac{13.7}{4.8}\right)$

B. $\arccos\left(\dfrac{13.7}{4.8}\right)$

C. $\arctan\left(\dfrac{13.7}{4.8}\right)$

D. $\arcsin\left(\dfrac{4.8}{13.7}\right)$

E. $\arccos\left(\dfrac{4.8}{13.7}\right)$

26. Tim sells notebooks. The number of notebooks, s, that Tim sells in 1 year depends on the price, p, of the notebook for that year, in dollars per notebook. The equation $s = 64 - 4p$, where $0 < p \leq 16$, gives the relationship between s and p. Tim's revenue is the money collected from selling s notebooks. The maximum revenue Tim can make from selling notebooks in 1 year occurs at what price per notebook?

F. $8

G. $9

H. $10

J. $12

K. $14

Advanced Mathematics Mixed Problem Set 3
Advanced Problem Set 16

27. A geometric sequence is a sequence of numbers in which each term is multiplied by a constant to obtain the following term. What is the 4th term in the geometric sequence with first 3 terms 9, 12 and 16?

   A. 21
   B. $\dfrac{16}{3}$
   C. $\dfrac{32}{3}$
   D. $\dfrac{64}{3}$
   E. 32

28. If $0 \leq \theta < 2\pi$ and $\sin^2 \theta + \sin \theta = \dfrac{3}{4}$, what is $\cos \theta$?

   F. $\dfrac{\sqrt{3}}{2}$
   G. $\sqrt{2}$
   H. $\dfrac{\pi}{6}$
   J. $\dfrac{1}{2}$
   K. 0

29. The system of equations below has multiple solutions, all of which satisfy the equation $y = \dfrac{1}{2}(x) - \dfrac{2}{3}$. If it can be determined, what is the value of c?

$$6x - 12y = 8$$
$$9x + cy = 12$$

   A. -18
   B. -12
   C. 4
   D. 9
   E. Cannot be determined

Advanced Mathematics Mixed Problem Set 3
Advanced Problem Set 16

30. As shown in the figure below, J is the center of the circle, and right triangle $\triangle GHJ$ intersects the circle at points K and I. Point I is the midpoint of $\overline{JH}$, which is 16 feet long. The shaded region inside the circle and outside the triangle has an area of 56π square feet. What is the measure of $\angle G$?

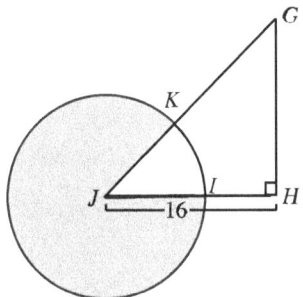

- **F.** $32°$
- **G.** $36°$
- **H.** $42°$
- **J.** $45°$
- **K.** $48°$

31. Depreciation can be modeled by the formula $V = I(1 - r)^t$, where I is the cars initial purchase price, r is the car's constant annual rate of decrease in value, expressed as a decimal: and V is the car's dollar value at the end of t years. A used car with a purchase price of $60,000 has a constant annual rate of decrease in value of 0.3. According to the model, what is the value of the car, to the nearest dollar, at the end of 4 years?

- **A.** $8,076
- **B.** $10,085
- **C.** $14,406
- **D.** $20,580
- **E.** $24,576

Advanced Mathematics Mixed Problem Set 3
Advanced Problem Set 16

32. Depreciation can be modeled by the formula $V = I(1 - r)^t$, where I is the cars initial purchase price, r is the car's constant annual rate of decrease in value, expressed as a decimal: and V is the car's dollar value at the end of t years.
    A used car decreased in value by 14% over 2 years. The car's initial price was $20,000. At what rate did the value of the car decrease, to the nearest hundredth?

    F. 0.057

    G. 0.073

    H. 0.078

    J. .860

    K. .927

33. Which of the following number line graphs represents the solution set to the equation $x^2 + 6 = 10$?

    A.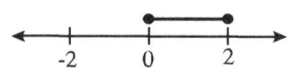

    B.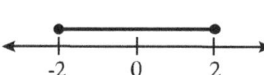

    C.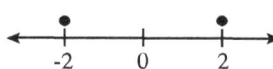

    D.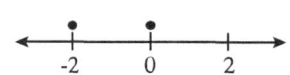

    E.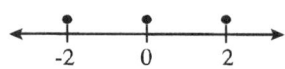

Advanced Mathematics Mixed Problem Set 3
Advanced Problem Set 16

34. This is the graph of $y = f(x)$
Which is the graph of $y = |f(x)|$?

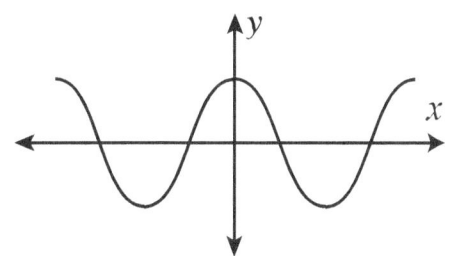

F.

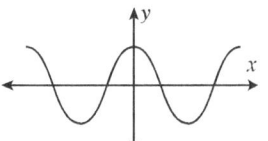

G.

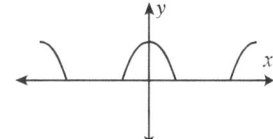

H.

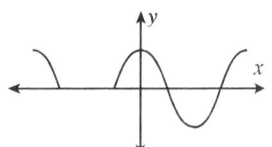

J.

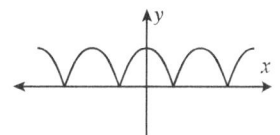

K.

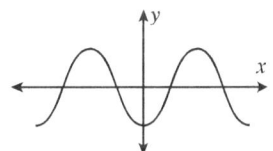

DO YOUR FIGURING HERE

35. The determinant of the matrix
$\begin{bmatrix} e & f \\ g & h \end{bmatrix}$ is $eh - fg$.

The determinant of the following array is 0.

$\begin{bmatrix} (x-4) & 11 \\ 4 & (x+3) \end{bmatrix}$

What are the possible values of x?

A. -8 and 7

B. -7 and 8

C. -4 and 3

D. -3 and 4

E. $\dfrac{1}{2} - \dfrac{\sqrt{124}}{2}$ and $\dfrac{1}{2} + \dfrac{\sqrt{124}}{2}$

Advanced Mathematics Mixed Problem Set 3
Advanced Problem Set 16

36. Given that

    $$c \begin{bmatrix} 6 & 8 \\ 3 & 4 \end{bmatrix} = \begin{bmatrix} d & 52 \\ e & f \end{bmatrix}$$

    for some real number c, what is $d + f$.

    F. $\dfrac{13}{2}$

    G. 52

    H. 26

    J. 39

    K. 65

Advanced Mathematics Mixed Problem Set 3
Advanced Problem Set 16

Answer Key

#	Answer	Frequency	Difficulty
1	E	average	2
2	K	average	1
3	D	popular	3
4	J	rare	3
5	C	popular	3
6	G	popular	3
7	A	rare	2
8	J	popular	1
9	B	popular	3
10	G	popular	1
11	A	popular	2
12	G	popular	1
13	C	popular	1
14	F	average	2
15	B	average	3
16	G	popular	2
17	D	popular	2
18	F	popular	3
19	B	average	3
20	J	popular	2
21	C	popular	1
22	J	popular	2
23	A	popular	3
24	H	popular	2
25	E	average	1
26	F	popular	3
27	D	average	1
28	F	popular	3
29	A	popular	1
30	J	average	1
31	C	popular	3
32	G	popular	3
33	C	popular	3
34	J	average	3
35	B	average	3
36	K	average	3

Advanced Mathematics Mixed Problem Set 4
Advanced Problem Set 17

1. For all x such that $\tan x \neq 0$, the expression $\dfrac{\csc^2 x \cos x}{\cot x}$ is equivalent to which of the following?

    A. 1

    B. $\sin x$

    C. $\sin^3 x$

    D. $\csc x$

    E. $\csc x \cot^2 x$

2. For such x such that $0 < x < \dfrac{\pi}{2}$, the expression $\dfrac{\sqrt{1-\sin^2 x}}{\cos x} - \dfrac{\sqrt{1-\cos^2 x}}{\sin x}$ is equal to:

    F. 0

    G. 1

    H. 2

    J. $\tan^2 x$

    K. $\sec 2x$

3. In $\triangle ABC$, the measure of $\angle A$ is $90°$, the measure of $\angle C$ is θ, $AB = 16$ units, and $\tan \theta = \dfrac{2}{3}$. What is the area of $\triangle ABC$, in square units?

    A. 15

    B. 50

    C. 75

    D. 150

    E. 192

4. Angle A has a measure of $\dfrac{41\pi}{4}$ radians. Angle A and Angle B are coterminal. Angle B could have which of the following measures?

    F. $4°$

    G. $30°$

    H. $45°$

    J. $60°$

    K. $180°$

Advanced Mathematics Mixed Problem Set 4
Advanced Problem Set 17

5. In trigonometry, an angle of $\dfrac{-8\pi}{5}$ radians has the same sine and cosine as an angle that has which of the following degree measures?

   A. $24°$
   B. $30°$
   C. $36°$
   D. $72°$
   E. $288°$

6. Four circles have the same center. Their radii measure 2, 4, 6 and 8 inches respectively. If a point is chosen at random in the interior of the largest circle, what is the probability that the point is also in the interior of the smallest circle?

   F. $\dfrac{1}{3}$
   G. $\dfrac{1}{4}$
   H. $\dfrac{1}{9}$
   J. $\dfrac{1}{12}$
   K. $\dfrac{1}{16}$

7. Which of the following is an equation of the circle with center at $(-4, 2)$ and a radius of 6 coordinate units in the standard (x, y) coordinate plane?

   A. $x^2 + y^2 + 8x - 4y = 6$
   B. $x^2 + y^2 + 8x - 4y = 16$
   C. $x^2 + y^2 + 8x - 4y = 36$
   D. $x^2 + y^2 - 8x + 4y = 16$
   E. $x^2 + y^2 - 8x + 4y = 36$

Advanced Mathematics Mixed Problem Set 4
Advanced Problem Set 17

8. What is the 164th digit to the right of the decimal point in the repeating decimal $0.\overline{762418}$?

   F. 1
   G. 2
   H. 4
   J. 6
   K. 7

9. The height that a certain ball rebounds after it hits the ground is directly proportional to the height from which it falls. When the ball falls from an initial height of 330 cm, it hits the ground for the first time and rebounds to a height of 240 cm. Arranged in order starting with the first bounce, the heights, in centimeters, that the ball rebounds form a sequence. Which of the following characterizes this sequence?

   A. Arithmetic with common difference -90
   B. Arithmetic with common difference 90
   C. Geometric with common ratio $\frac{11}{8}$
   D. Geometric with common ratio $\frac{8}{11}$
   E. Neither arithmetic nor geometric

10. There is an arithmetic sequence with a positive common difference, in which the sum of the first 3 terms of the sequence is 162. Which of the following values cannot be the first term of the arithmetic sequence?

    F. 30
    G. 34
    H. 45
    J. 52
    K. 57

Advanced Mathematics Mixed Problem Set 4
Advanced Problem Set 17

11. What is the real value of x in the equation $\log_3 45 - \log_3 5 = \log_4 x$?

    A. 3
    B. 9
    C. 16
    D. 25
    E. 125

12. For all $x > 2$, $\log(x^2 + x - 6) + \log(x) =$

    F. $\log(x^2 + 2x - 6)$
    G. $\log(x^2 - 6)$
    H. $\log\left(\dfrac{x^2 + x - 6}{x}\right)$
    J. $\log\left(\dfrac{x}{x^2 + x - 6}\right)$
    K. $\log(x^3 + x^2 - 6x)$

13. The solution set for the equation $13^{x^2+4} = 1$ contains:

    A. 2 imaginary numbers
    B. 2 positive real numbers
    C. 1 negative and 1 positive real number
    D. 1 negative real number only
    E. 1 real number, which is 0

14. Consider the functions $g(x) = \sqrt{x}$ and $h(x) = 4x + c$. In the standard (x, y) coordinate plane, $y = g(h(x))$ passes through $(5, 5)$. What is the value of c?

    F. 5
    G. -5
    H. 15
    J. -15
    K. -25

Advanced Mathematics Mixed Problem Set 4
Advanced Problem Set 17

15. In the standard (x, y) coordinate plane, when $c \neq 0$ and $d \neq 0$, the graph of $f(x) = \dfrac{6x + c}{2x + d}$ has a horizontal asymptote at:

    A. $y = 3$
    B. $y = 6$
    C. $y = c$
    D. $y = \dfrac{c}{d}$
    E. $y = d$

16. For what integer(s) h are both solutions of the equation $x^2 + hx + 13 = 0$ positive integers?

    F. -14
    G. -12
    H. -2
    J. 1
    K. 16

17. A number is increased by 35% and the resulting number is then decreased by 20%. The final number is what percent of the original number?

    A. 90%
    B. 98%
    C. 100%
    D. 108%
    E. 116%

18. What are the real solutions to the equation $|x|^2 - 2|x| - 8 = 0$?

    F. ±4
    G. ±2
    H. 2 and 4
    J. −2 and −4
    K. ±2 and ±4

DO YOUR FIGURING HERE

Advanced Mathematics Mixed Problem Set 4
Advanced Problem Set 17

19. Jeremiah's weight does not vary more than 12 pounds from p pounds. Which of the following inequalities gives the range of Jeremiah's weight, w, in pounds?

    A. $|w - p| \leq 12$

    B. $|w + p| \leq 12$

    C. $|w - 12| < p$

    D. $w - 12 < p$

    E. $w + 5 < p$

20. The solution of the system of equations below is the set of all (x, y) such that $3x + 2y = 12$. What is the value of h?
    $$27x + 18y = 108$$
    $$9x - hy = -6h$$

    F. 9

    G. 6

    H. 3

    J. -6

    K. -9

21. The length of the shorter side of rectangle $EFGH$ is 5 inches less than the length, L, of the longer side. The length of the longer side of rectangle $LMNO$, which is similar to $EFGH$, is 8L inches. In terms of L, what is the length of the shorter side of $LMNO$?

    A. $L - 40$

    B. $L + 5$

    C. $8L + 5$

    D. $8L - 5$

    E. $8L - 40$

Advanced Mathematics Mixed Problem Set 4
Advanced Problem Set 17

22. Given $f(x) = \sqrt[4]{x-5}$, which of the following expressions is equal to $f^{-1}(x)$ for all real numbers x?

    F. $\sqrt[4]{x-5}$

    G. $\sqrt[4]{x+5}$

    H. $-\sqrt[4]{x+5}$

    J. $x^4 + 5$

    K. $(x+5)^4$

23. In teaching a lesson on the concept of fourths, Ms. Saltzer uses a divide-and-set-aside procedure. She starts with a certain number of colored disks, divides them into 4 equal groups, and sets 1 group aside to illustrate $\dfrac{1}{4}$. She repeats the procedure by taking the disks she had NOT set aside, dividing them into 4 equal groups, and setting 1 of these groups aside. If Ms. Saltzer wants to be able to complete the divide-and-set-aside procedure at least 4 times without breaking any of the disks into pieces, which of the following is the minimum number of colored disks she can start with?

    A. 4
    B. 16
    C. 32
    D. 64
    E. 256

24. For a project in class, Stephen is making a tablecloth for a circular table 4 feet in diameter. The finished tablecloth needs to hang down 5 inches over the edge of the table all the way around. To finish the edge of the tablecloth, Stephen will fold under and sew down 1 inch of the material all around the edge. What is the shortest length of fabric, in inches, Stephen could use to make the tablecloth without putting any separate pieces of fabric together?

    F. 16
    G. 21
    H. 48
    J. 58
    K. 60

Advanced Mathematics Mixed Problem Set 4
Advanced Problem Set 17

25. Each contestant at a math competition starts with 40 points. A contestant earns 15 points for each question answered correctly and loses 5 points for each question answered incorrectly. Caroline answered three times as many questions correctly as incorrectly. She finished with 280 points. How many questions did Caroline answer correctly?

    A. 24
    B. 18
    C. 12
    D. 8
    E. 6

26. Andrew drove from Oretown to Blare City, a distance of 120 miles. From Blare City he drove on to Jeffersonville, and then drove back to Oretown. The ratio of Andrew's driving times on the first, second, and third segments of the trip, respectively, was 6:4:9, and he drove at the same average speed on each segment. What was Andrew's total driving distance, in miles, for the 3 segments of the trip?

    F. 120
    G. 180
    H. 300
    J. 380
    K. 460

27. Which of the following statements is a logical conclusion from the 3 true statements given below?

    All leeps are grools
    All parps are grools
    All krins are leeps

    A. No parps are grools
    B. No parps are krins
    C. All krins are grools
    D. All parps are leeps
    E. All leeps are parps

DO YOUR FIGURING HERE

Advanced Mathematics Mixed Problem Set 4
Advanced Problem Set 17

Answer Key

#	Answer	Frequency	Difficulty
1	D	popular	4
2	F	popular	5
3	E	popular	4
4	H	popular	3
5	D	popular	3
6	K	average	2
7	B	rare	4
8	J	average	3
9	D	average	4
10	K	average	4
11	C	average	4
12	K	average	3
13	A	average	3
14	F	popular	3
15	A	average	4
16	F	popular	3
17	D	popular	3
18	F	popular	2
19	A	popular	3
20	J	popular	3
21	E	average	3
22	J	popular	4
23	E	popular	4
24	K	popular	2
25	B	popular	3
26	J	popular	3
27	C	rare	3